Measuring America

ANDRO LINKLATER, son of the novelist Eric Linklater, was born in Scotland in 1944 and brought up there. After studying history at Oxford University, he worked for several years in the United States. For the last twenty years he has been a full-time writer and journalist, travelling extensively throughout the world. His books include an award-winning biography of Compton Mackenzie, a history of the Black Watch and an account of his experiences among the headhunters of Sarawak.

Andro Linklater is married to a photographer, Marie-Louise Avery, and lives in Kent.

'Wonderful and fascinating . . . This clever and eminently readable book explains why and how land came to be owned in America in the first place. It is a huge topic, and it makes for a hugely satisfying read. Of all the slender books to have occupied this amply filled new genre of publishing, Linklater's is by far the most learned, satisfying, and fun. It deserves to be a classic, plunged into the reader's consciousness as firmly as the iron spikes or the witness trees at the edges of the maps it so splendidly describes'

SIMON WINCHESTER, *Boston Globe*

'A *tour de force*. The extent to which [Linklater] succeeds in his ambition of getting the reader to take a fresh look at America is breathtaking'   RICHARD DELEVAN, *Sunday Business Post*

'A subject of extraordinary interest and relevance, which has lent itself to a fine historical survey' ROGER HUTCHINSON, *Scotsman*

'What an ⟨engag⟩ ⟨...⟩ ghly fascinating story to te⟨ll, se⟨...⟩ ⟨...⟩ ⟨...⟩e, and no small element o⟨...⟩ sur⟨...⟩ ⟨...⟩e adventure of ideas, oug⟨ht to⟩ ⟨...⟩e first page'
⟨...⟩ of *John Adams*

'A luminous and ⟨...⟩ ⟨...⟩ ⟨...⟩f ideas and the passions and foibles of men'
JAY WINK, author of *April 1865: The Month that Saved America*

MEASURING AMERICA

*How the United States was Shaped
by the Greatest Land Sale
in History*

ANDRO LINKLATER

HarperCollins*Publishers*

HarperCollins*Publishers*
77–85 Fulham Palace Road,
Hammersmith, London w6 8jb

The HarperCollins website address is:
www.**fire**and**water**.com

This paperback edition 2003
1 3 5 7 9 8 6 4 2

First published in Great Britain by
HarperCollins*Publishers* 2002

Copyright © Andro Linklater 2002

ISBN 0 00 710888 5

Set in PostScript Linotype New Baskerville

Printed and bound in Great Britain by
Clays Ltd, St Ives plc

CONTENTS

ILLUSTRATIONS

The young George Washington holding the surveyor's
essential tools, Gunter's chain and a circumferentor or
primitive theodolite.

Gunter's chain. (© *Science Museum/Science and Society Picture
Library*)

Gunter's quadrant or quarter circle. (© *Science Museum/Science
and Society Picture Library*)

A theodolite made in 1737 by J. Sisson. (© *National Maritime
Museum*)

A German representation in 1583 of the basic land measure,
the rod, pole or perch, as it was becoming standardised.

A sixteenth-century scene of two surveyors using cross-staffs to
measure distance.

Thomas Jefferson (1743–1826). Third president, author of
the Declaration of Independence, and instigator of the
squares that spread across the nation. (© *Hulton/Archive*)

The Connecticut Settlers Entering the Western Reserve by Howard
Pyle (1853–1911). (*Private collection/Bridgeman Art Library*)

Rufus Putnam (1738–1824).

The Rev. Manasseh Cutler (1742–1823).

William A. Burt (1792–1858), Michigan surveyor and
inventor of the solar compass. (© *National Portrait Gallery,
Smithsonian Institute; gift of Mrs Philip Burt Fisher*)

Ferdinand Rudolph Hassler (1770–1843), the United States's
master measurer.

The Paris–Greenwich triangulation. The point where the
Cassinis' triangles meet William Roy's. (© *National Maritime
Museum*)

A poster dated about 1800 attempting to reconcile the
French to the joys of metrication.

First Furrow by Olaf Carl Seltzer (1877–1975). (*Thomas Gilcrease Institute, USA*)

An early-nineteenth-century view of Washington, DC. (*View of Washington, pub. by E. Sachse & Co., 1852, by American School Library of Congress, Washington D.C., USA/Bridgeman Art Library*)

Joseph Dombey (1742–1794), sculpted by Jean-Antoine Houdon. (© *Bibliothèque Centrale du Musée National d'Histoire Naturelle, 2002*)

The squares of the public land survey come up against the long lots of French settlers on the Red River in Louisiana.

An 1850 map of San Francisco, a neat grid of squares laid across precipitous slopes and rugged hills. (© *CORBIS*)

Western San Francisco, where the grid that Jasper O'Farrell drew spectacularly over the distant hills of the city spreads neatly towards the Pacific. (© *Skyscan/Jim Wark*)

The Illinois Central railroad's standard town plan, used to create hundreds of identical towns across Illinois.

A Kansas Land Office in 1874, with a potential purchaser being shown a pre-surveyed plot.

Salt Lake City in 1870.

The Oklahoma Land Rush. The start of a race to stake claims to land that had belonged to American Indians and was about to become American property. (© *Bettmann/CORBIS*)

A survey team near what would become the United States's first national park at Yellowstone in the Rockies. (© *CORBIS*)

The 1871 team under John W. Powell that surveyed the Colorado river valley. (© *CORBIS*)

ACKNOWLEDGEMENTS

Triangulation, the surveyor's standby, created the idea for this book. The first point was a chance reference to Joseph Dombey, the unlucky messenger sent by France to the United States in 1794 with a copy of the prototype kilogram and metre. The second was the vivid memory of a flight from Los Angeles to New York on a clear winter's day, which revealed a chequered land that reached from east of the Rockies to Pennsylvania. The third was a chapter in John Stilgoe's *The Common Landscape of America* describing the effect of the public land survey on the appearance of the nation's countryside.

To fill in the ground between those distant points I relied enormously on the skill, experience and generosity of many people on both sides of the Atlantic. In roughly chronological order, I should like to express my gratitude to Malcolm Draper for kindly sharing his vast experience of surveying, demonstrating the use of a theodolite, and at a later stage for reading part of the manuscript; to Stephen Booth, editor of *Geomatic World*, for his encouragement and supply of vital information; to James Kavanagh of the Geomatics Faculty at the Royal Institution of Chartered Surveyors, for his friendship and for introducing me to the Institution's library; to Richard Johnson, my publisher at HarperCollins, who not only took the idea seriously but told others about it; to George Gibson of Walker and Company, New York, who shared his enthusiasm; to Lyn Cole, who undertook research in France with undaunted pertinacity; to my wife, Marie-Louise, for driving with me through five thousand miles of American back-roads with only one argument, and for opening my eyes to innumerable aspects of the passing landscape that I should otherwise have missed; to Holly Iverson in Fargo,

North Dakota, who shared her family history of life in the state; to Catherine Renschler of Adams County, Nebraska, for her personal and historical insights into pioneer life on the prairies; to Tom Schmiedeler and Barbara Jarvis in Lawrence, Kansas, for their kind hospitality; to Lance Bishop, chief of Geographical Services at the Bureau of Land Management in Sacramento, California, for giving generously of his time and resources; to Ed Patton, heir to the tradition of Thomas Hutchins, who showed me what the public land survey was all about; to Gerard Iannelli of the Metric Program for his information about the attempt to introduce metric measurement to the United States; to Lawrence Brooks for advice on quit rents; to Penry Williams for reading the chapter on Tudor history; and to Robert Lacey for his attentive editing. 'The Gift Outright', from *The Poetry of Robert Frost*, edited by Edward Connery Lathem, © The Estate of Robert Frost, published by Jonathan Cape, is reproduced by permission of The Random House Group Limited.

Among institutional sources of information, I should especially like to thank for their professional help and the use of their resources the staffs at the London Library; the British Library; the Royal Institution of Chartered Surveyors' library; the Campus Martius Museum at Marietta, Ohio; the Michigan Museum of Mining; the State Archives at the Michigan State Museum; the Marquette County Historical Society; the North Dakota Institute for Regional Studies; the Adams County Historical Society in Nebraska; the archives of the Bureau of Land Management, Sacramento, California; the Library of Congress; the library and museum of the National Institute of Standards and Technology in Gaithersburg, Maryland; the New York Public Library; and, for its miraculous speed at accessing information on the web, Google.

Responsibility for errors of omission and commission, on the other hand, is due entirely to me.

ANDRO LINKLATER
Markbeech, Kent
January 2002

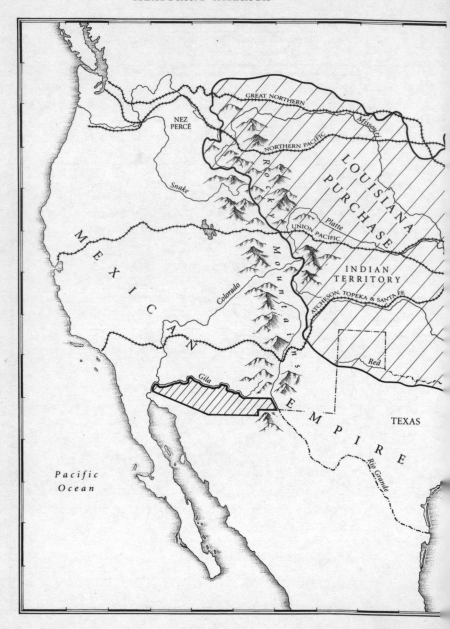

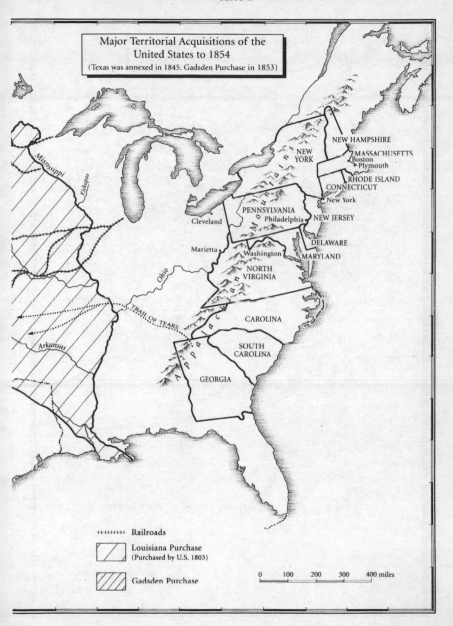

Major Territorial Acquisitions of the
United States to 1854
(Texas was annexed in 1845. Gadsden Purchase in 1853)

Mississippi

Kickapoo

NEW HAMPSHIRE

NEW YORK

MASSACHUSETTS
Boston
Plymouth

RHODE ISLAND

CONNECTICUT

New York

Cleveland

PENNSYLVANIA
Philadelphia

NEW JERSEY

DELAWARE

Marietta

Washington

MARYLAND

Ohio

NORTH
VIRGINIA

TRAIL OF TEARS

Arkansas

CAROLINA

SOUTH
CAROLINA

Appalachian

GEORGIA

⊦⊦⊦⊦⊦⊦⊦⊦ Railroads

Louisiana Purchase
(Purchased by U.S. 1803)

Gadsden Purchase

0   100   200   300   400 miles

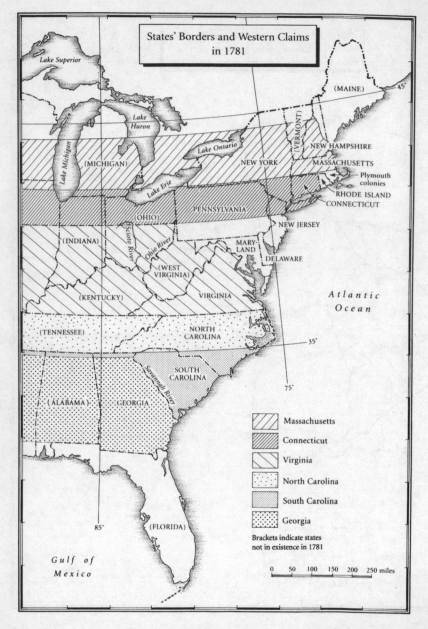

States' Borders and Western Claims
in 1781

Lake Superior

Lake Huron

Lake Michigan

(MICHIGAN)

Lake Ontario

Lake Erie

(OHIO)

(INDIANA)

Scioto River

Ohio River

(WEST VIRGINIA)

(KENTUCKY)

(MAINE)

(VERMONT)

NEW HAMPSHIRE

NEW YORK

MASSACHUSETTS

Plymouth colonies

RHODE ISLAND

CONNECTICUT

PENNSYLVANIA

NEW JERSEY

MARY-LAND

DELAWARE

VIRGINIA

Atlantic Ocean

(TENNESSEE)

NORTH CAROLINA

Savannah River

SOUTH CAROLINA

(ALABAMA)

GEORGIA

45°

35°

75°

85°

(FLORIDA)

Gulf of Mexico

	Massachusetts
	Connecticut
	Virginia
	North Carolina
	South Carolina
	Georgia

Brackets indicate states
not in existence in 1781

0   50   100   150   200   250 miles

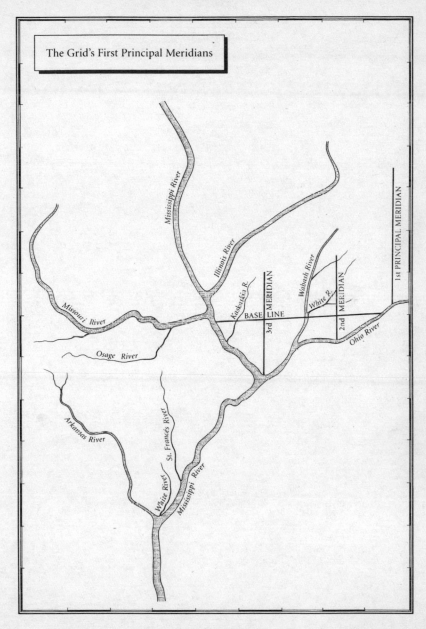

The Grid's First Principal Meridians

INTRODUCTION

In every country with an industrial history there are towns like East Liverpool, Ohio. They were built close to the coalfields that provided their energy and on the banks of large rivers that carried away their heavy products. On the Clyde in Scotland, the Tees in England, the Ruhr in Germany, mighty structures of iron and steel were smelted, beaten and annealed to make the skeleton for the first modern society, and the towns grew rich and confident in a grey haze of fumes and steam and effluent. Today their colours are rusty metal and faded red brick, the air is clear, the atmosphere uncertain, and in the shabby streets regeneration is pitted against boredom and drugs.

Clay was the material that made East Liverpool wealthy, Pennsylvania coal fuelled the furnaces, and the Ohio river carried away enough plates and cups to let the inhabitants boast that the town was 'the pottery capital of the world'. Where once dozens of factories and chemical works lined the banks, the only reminder now of East Liverpool's industrial past is a solitary chimney stack pumping heavy coils of smoke into the sky. The barges still come up the Ohio river carrying minerals and fertilisers, but mostly for use across the river in the nearby Pittsburgh area. They unload at depots like that belonging to the S.H. Bell Company, whose functional warehouses and concrete docks annually handle thousands of tons of steel and copper at the upriver end of town.

The place could hardly be more anonymous. Above the Bell Company's dock, Pennsylvania Route 68 invisibly changes to

Ohio Route 38, and trees half-hide some signs by the roadside. Even someone familiar with the historical significance of this particular spot, who has travelled several thousand miles to find it, and whose eyes are flickering wildly from the narrow blacktop to the nondescript verge between road and river, can drive a couple of hundred yards past it before hitting the brakes.

The language of the signs is equally undemonstrative. A stone marker carries a plaque headed 'The Point of Beginning' which reads, '1112 feet south of this spot was the point of beginning for surveying the public lands of the United States. There on September 30 1785, Thomas Hutchins, first Geographer of the United States, began the Geographer's Line of the Seven Ranges.'

There is nothing else to suggest that it was here that the United States began to take physical shape, nothing to indicate that from here a grid was laid out across the land that would stretch west to the Pacific Ocean, and north to Canada, and south to the Mexican border, and would cover over three million square miles, and would create a structure of land ownership unique in history, and would provide the invisible web that supported the legend of the frontier with its covered wagons and cowboys, and its farmers and goldminers, and would insidiously permeate its formation into the unconscious mind of every American who ever owned a square yard of soil.

It is hilly country, covered in the same oak, dogwood and hickory that Hutchins saw, and in the bright light of a September afternoon it rises high above the broad river in crimson and copper waves. 'For the distance of 46 chains and 86 links West,' Hutchins wrote in his first description of the territory, 'the Land is remarkably rich with a deep, black Mould, free from Stone.' He was Robinson Crusoe, landed in an uncharted wilderness, and his purpose was something close to magic – to measure it and map it and turn it into property. It had been lived in by the Delaware and passed through by the

Miami and occupied by the Iroquois, but no one had ever owned it. No one had ever thought of owning it. The idea of one person *owning* land did not yet exist on the west bank of the Ohio.

The wand that would make this magic possible was there in the first sentence of Hutchins' report. The land would be measured in chains and links. In most circumstances, a chain imprisons; here it released. What it released from the billowing, uncharted land was a single element – a distance of twenty-two yards. That was the length of the chain. Repeated often enough, added, squared and multiplied, the measurement gave a value to the land that could be computed in money terms.

What began here on the banks of the Ohio river was not just a survey. The real significance of the spot now covered by the Bell Company's concrete dock is that this was where the most potent idea in economic history – that land might be owned, like a horse or a house – was first released into the western wilderness and encouraged to spread across the land mass of the United States. But in his poem 'The Gift Outright', Robert Frost caught at a thought more powerful yet. The earth has its own magic, and those who seek to possess it run the risk of being possessed by it. It was the desire to own this particular land, Frost mused, that made its owners American.

These were the great forces that Captain Thomas Hutchins, Geographer to the United States, set in motion when he first unrolled the loops of his chain at the Point of the Beginning.

ONE

The Invention of Property

THE IMPOSING LIBRARY of the Royal Institution of Chartered Surveyors in London is strategically situated. In one direction its tall windows look over the street to Whitehall, where the Tudor and Stuart sovereigns ruled in the sixteenth and seventeenth centuries, and in another they gaze across Parliament Square towards the House of Commons, power-base of the rising class of landed gentry who during those two centuries challenged the royal authority. It is just possible to imagine the atmosphere of righteous indignation and pervading apprehension which accompanied the struggle between the two, but in the small, 450-year-old, leather-bound books kept in the Institution's library, the reality that gave rise to the battles remains vividly alive.

In the earliest, such as Master Fitzherbert's *The Art of Husbandry*, published in 1523, a surveyor still fills his original, feudal role as the executive officer of a landed nobleman. His duty is simply to oversee (the word 'surveyor' is derived from the French *sur* = over and *voir* = see) the estate. He is to walk over the land, noting the 'buttes and bounds' of the tenants' holdings, and then to assist in drawing up the official record or court roll of what duties they owed. A model report, Fitzherbert suggests, might run like this: the land of a particular tenant 'lyeth between the mill on the north side, and the South Field on the south, butteth upon the highway, and

1

conteyneth xii [twelve] perches [a perch, like a rod, equals 16½ feet] and x [ten] fote [feet] in bredthe by the hyway, and ix [nine] perches in length, and payeth . . . two hennes at Christmas and two capons at Easter'.

To 'butt' upon something was to encounter or meet it; the alternative word was 'mete'. This ancient method of surveying, which identified the boundary of an estate by the points where it met other boundaries or visible objects, thus became known as 'metes and bounds'. Under that title it was to cross the Atlantic to the colonies of Virginia and Carolina and thence to Tennessee and Kentucky, to the confusion of landowners and the enrichment of lawyers.

Even in 1523, English landlords were engaged in a practice that was to transform the feudal order. There were infinite variations in feudalism, but at its heart was the principle that the land was the state, and only the head of state could own it outright. The dukes and barons, the king's tenants-in-chief, technically held their broad acres *of* the Crown in return for the dues or service they paid; their vassals held their narrower farms *from* the great lords in return for rent or service; and so on down to the villeins, who had no land at all, but exchanged goods, service or rent for the right to work it. The feudal principle applied equally to the American colonies, whether they were founded by commercial concerns like the Virginia Company, or individual proprietors like William Penn. Every charter authorising the foundation of a colony, from Virginia in 1606 to Georgia in 1732, declared that the land was held of the King 'as of his mannor of East Greenwiche in the county of Kent, in free and common soccage' – a term which in fact imposed few obligations, but recognised the feudal framework governing land ownership on either side of the Atlantic. What the sixteenth-century manuals inadvertently reveal, as they detail the surveyor's duties, is how that order was subverted from within.

Under the old system, tenants farmed narrow strips or rigs

of land, often widely separated so that good and poor soil was distributed evenly among those who actually worked the land. For centuries, land-users had attempted to consolidate the strips into single compact fields which could be 'enclosed' by a fence or hedge so that crops were not trodden down or herds scattered, but the pattern remained fundamentally intact. Now, however, a period of savage inflation occurred, and in the early sixteenth century every lord and tenant was trying to squeeze the maximum profit from the land. Repeatedly Fitzherbert stresses the need for the surveyor to realise that enclosed land was more valuable than the strips and common pasture because it could be made more productive. The pressure for change is unmistakable, yet essentially the old values are still in place.

Then in 1534 comes the publication of *The Boke named the Governour*, by Sir Thomas Elyot, which gives advice to 'governors', whether of kingdoms or estates, on how to run their 'dominions'. An essential first step according to Elyot is to draw a map or 'figure' of the estate so that the governor knows what it consists of, or as he puts it, 'in visiting his own dominions, he shall set them out in figure, in such wise that his eye shall appear to him where he shall employ his study and treasure'. In the course of the sixteenth century, it became a habit of English landowners to have their estates and the surrounding countryside measured and then mapped. By 1609 John Norden could insist in the *Surveior's Dialogue* that 'the [map] rightly drawne by true information, describeth so lively an image of a Manor . . . as the Lord sitting in his chayre, may see what he hath, where and how it lyeth, and in whose use and occupation every particular is'.

There was a particular significance in the surveyor's new duty of mapmaking, because in that era only the rulers of states and cities made maps. A map was a political document. It not only described territory, but asserted ownership of it. From 1549 a map of Newfoundland and the North Atlantic seaboard of North America detailing Sebastian Cabot's discoveries used to hang in the Privy Gallery at Whitehall outside the royal council chamber, so that foreign ambassadors waiting to see the sovereign would know of England's claims overseas. When the Flemish cartographer Abraham Ortelius produced the first modern atlas in 1570, his *Theatrum orbis terrarum*, which included the freshly discovered territories of the New World and the newly explored Pacific and Indian Oceans, he took care to dedicate it to his sovereign, Philip II of Spain, and to ensure that Philip could find in it his own claims across the ocean.

Consequently, when the sixteenth-century English landowners ordered maps of their estates, they were making a very particular claim. For a long time almost no one but the English made such a claim. Surveying manuals were published in the German states, but there were hardly any estate maps until late in the seventeenth century. Sweden produced her first national map in the sixteenth century, but it was a hundred years later before noblemen began measuring and mapping their estates. In sixteenth-century France, the Jesuits taught maths and all the theory needed by a surveyor, but, as the distinguished historian Marc Bloch noted, no plats or *plans parcellaires* were drawn before 1650. The first Spanish maps appeared as early as 1508, but no Spanish lords showed any interest in measuring their lands for another two hundred years. Only in the economically sophisticated Netherlands, where the mathematician Gemma Frisius wrote the first manual on mapmaking, *A Method of Delineating Places*, in 1533, were farms, especially those close to cities, measured and mapped, yet even there the aristocrats' landed estates remained feudal. But in England, Henry VIII collected so many estate maps that an inventory of his possessions at his death in 1547 showed he had 'a black coffer covered with fustian of Naples [which was] full of plattes'.

The significance was unmissable. When English landowners

commissioned surveyors to measure out their estates and make maps or plats of them, they were asserting a form of ownership that until then only rulers and governors could make.

If there is a single date when that idea of land as private property can be said to have taken hold, it is 1538. In that year a tiny volume was published with a long title which begins, '*This boke sheweth the maner of measurynge of all maner of lande . . .*'. In it, the author, Sir Richard Benese, described for the first time in English how to calculate the area of a field or an entire estate. He was probably borrowing his methods from Frisius, but his values were purely English. Noting that sellers tend to exaggerate the size of a property while buyers are inclined to underestimate it, he advises the surveyor to approach the task in a careful and methodical manner.

'When ye shall measure a piece of any land ye shall go about the boundes of it once or twice, and [then] consider well by viewing it whether ye may measure it in one parcel wholly altogether or else in two or many parcels.' Measuring in 'many parcels', he explains, is necessary when the field is an uneven, irregular shape; by dividing it up into smaller, regular shapes like squares and oblongs and triangles it becomes easy to calculate accurately the total area. The distances are to be carefully measured with a rod or pole, precisely 16½ feet long, or a cord. And finally the surveyor is to describe the area in words, and to draw a plat showing its shape and extent.

Like the maps, this interest in exact measurement is new. Before then, what mattered was how much land would yield, not its size. When William the Conqueror instituted the great survey of England in 1086, known as the Domesday Book, his commissioners noted the dimensions of estates in units like virgates and hides, which varied according to the richness of the soil: a virgate was enough land for a single person to live on, a hide enough to support a family; consequently the size shrank when measuring fertile land, and expanded in poor, upland territory. Other Domesday units like the acre and the carrucate were equally flexible, but so long as land was held in exchange for services, the number of people it could feed and so make available to render those services was more important than its exact area. Accurate measurement became important in 1538 because, beginning in that year, a gigantic swathe of England – almost half a million acres – was suddenly put on sale for cash.

The greatest real-estate sale in England's history occurred after Henry VIII dissolved a total of almost four hundred monasteries which had been acquiring land for centuries. He justified his action on the grounds that these houses of prayer had grown depraved and corrupt, but tales of drunken monks and lecherous nuns served to conceal a more mundane purpose – Henry needed money. On the monasteries' dissolution, all their land, including some of the best soil in England, automatically reverted to their feudal overlord, the king. These rich acres were then sold to wealthy merchants and nobles in order to pay for England's defences.

The sale of so much land for cash was a watershed. Although changes were already underway, with feudal services often commuted for rents paid in coin, and feudal estates frequently mortgaged and sold, up to that point the fundamental value of land remained in the number of people it supported. From now on the balance would shift increasingly to a new way of thinking. Prominent among the purchasers of Church property were land-hungry owners, like the Duke of Northumberland, who had been enclosing common pastures, but far more common were the small landlords who had done well from the rise in the market value of wool and corn, and who now chose to invest in monastery estates. In Norfolk, Sir Robert Southwell attracted attention because of the mighty pastures he carved out from common land for his fourteen flocks of sheep, each numbering around a thousand animals, but the Winthrop family who acquired and enclosed monastery land in nearby Suffolk almost escaped notice.

They and their surveyors knew that the monasteries' widely separated rigs and shares of common land would become more valuable once they were consolidated into fields. It was what later generations would term an investment opportunity. The old abbots and priors had understood land ownership to be part of a feudal exchange of rights for services. The new owners knew that it depended on money changing hands, and that to maximise profits the old ways had to be replaced.

'Jesu, sir, in the name of God what mean you thus extremely to handle us poor people?' a widow demanded of John Palmer, an enclosing landlord in Sussex who in the 1540s bought the monastic estate on which she lived, and evicted her from her cottage.

'Do ye not know that the King's grace hath put down all the houses of monks, friars and nuns?' Palmer retorted. 'Therefore now is the time come that we gentlemen will pull down the houses of such poor knaves as ye be.'

As enclosures and rising rents forced thousands of villeins and farm-labourers away from the manors that once supported them, protests kept the printing presses busy. Some represented the propaganda of conservative voices like Sir Thomas More, who memorably wrote of pastoralists like Southwell, 'Your sheep that were wont to be so meek and tame, and so small eaters, now as I hear say, be become so great devourers and so wild, that they eat up and swallow down the very men themselves.' Resentment also triggered popular uprisings in the north, east and west of England, and the government itself introduced Bills in Parliament against enclosure, though by the second half of the sixteenth century few were passed.

Recent research downplays the actual number of enclosures, but no one questions the enormous redistribution of land that occurred. Even G.R. Elton, the most sceptical of Tudor historians, accepted that it 'laid the foundation for that characteristic structure of landlord, leasehold farmer, and landless labourer which has marked the English countryside from that day to this'.

The hidden hand in this gigantic upheaval was provided by the survey and plat that recorded the new owner's estate as his property. The emphasis in Benese's book on exact measurement reflected the change in outlook. Once land was exchanged for cash, its ability to support people became less important than how much rent it could produce, and that depended largely on the size of the property. The units used to measure this could no longer vary; the method of surveying had to be reliable. The surveyor ceased to be a servant, and became an agent of change from a system grounded in medieval practice to one which generated money.

Some at least became uneasily aware of what they were doing. In the *Surveior's Dialogue,* John Norden specifically blamed the act of measuring itself for helping to destroy the old ways, and held surveyors responsible as 'the cause that men [lose] their Land: and sometimes they are abridged of such liberties as they have long used in Mannors: and customes are altered, broken, and sometimes perverted or taken away by your means'.

What the new class of landowners required of their surveyors above all was exactness, and the sudden increase in the number of manuals the last quarter of the sixteenth century testified to the urgency of their need. Before that, in 1551, Robert Recorde wrote a book called *Pathway to Knowledge* in praise of the accuracy that geometry offered surveyors, but warning of its potential for destruction:

> Survayers have cause to make muche of me.
> And so have all Lordes that landes do possesse:
> But Tennauntes I feare will like me the lesse.
> Yet do I not wrong, but measure all truely,
> And yelde the full right to everye man justely.
> Proportion Geometricall hath no man opprest,
> Yf anye bee wronged, I wishe it redrest.

It was against this background – an urgent and growing need for the accurate measurement of land – that Edmund Gunter devised his chain. Born in 1581 to a Welsh family, Gunter had been sent to Oxford University to be educated as a Church of England priest, but by the time he was ordained he had discovered that numbers were more inspiring to him than religion. In twelve years as a divinity student he preached just one sermon, and its reputation endured long after his death because, according to Oxford gossip, 'it was such a lamentable one'. What really interested him was the relationship of mathematics to the real world, and consequently he spent most of his time making instruments to illustrate the way in which numbers worked.

Ratios and proportions were his passion. He invented an early slide-rule, known as Gunter's scale, to demonstrate proportional connections between numbers, and worked out to seven places of decimals the logarithms for sine and cosine. The point at which this enthusiasm for numerical ratios touched upon concrete reality was trigonometry, which allowed mathematicians to calculate the length of two sides of a triangle, when only the third side and two angles were known; it also enabled the more expert surveyors to work out the distance between two objects without having to walk between them.

Since only the most basic instruments existed at that time, mathematicians and astronomers were expected to design their own. To demonstrate the solutions to problems in geometry, Gunter was constantly adapting and improving nautical instruments like the quadrant and cross-staff, which measured vertical angles between the sun and the horizon, or horizontal angles between towers, trees and churches. Indeed his enthusiasm for new gadgets cost him the best scientific job in the land. In 1620 the wealthy but earnest Sir Henry Savile put up money to fund Oxford University's first two science faculties, the chairs of Astronomy and Geometry. Gunter applied to become Professor of Geometry, but Savile was famous for distrusting clever people – 'Give me the plodding student,' he insisted drearily – and the candidate's behaviour annoyed him intensely. As was his habit, Gunter arrived with his sector and quadrant, and began demonstrating how they could be used to calculate the position of stars or the distance of churches, until Savile could stand it no longer. 'Doe you call this reading of Geometrie?' he burst out. 'This is mere showing of tricks, man!', and according to a contemporary account, 'dismisst him with scorne'.

Fortunately Gunter was supported by the Earl of Bridgewater, who did like brilliance, having grown up in a house where poets like Edmund Spenser and Ben Jonson were guests and where *Othello* was first performed. Since his father had inherited huge estates on the Welsh border and acquired valuable land north of London, the Earl was also even richer than Savile, and it seems probable that the surveyor's chain that Gunter designed in about 1607 was first used to measure the immense Bridgewater property.

Aided by aristocratic influence, Gunter was then appointed rector of the wealthy parish of St George's, Southwark, in London, and, in 1619, Professor of Astronomy at Gresham's College, London. However, both his congregation and his students were utterly neglected in favour of his scientific instruments. Like electronic devices today, these were sold with a book of incomprehensible instructions. Gunter at least had the excuse that few could understand his instructions because they were written in Latin. In 1623 the chorus of complaints persuaded him to produce a translation. 'I am at the last contented that it should come forth in English,' he wrote. 'Not that I think it worthy either of my labour or the publique view, but to satisfy their importunity who not understanding the Latin yet were at the charge to buy the instrument.'

The complete collection of Gunter's instructional books

issued in 1623 was called *The description and use of the sector, the cross-staffe and other instruments for such as are studious of mathematical practise.* By then Gunter must have known that the last bit of the title was nonsense. The reason the book had to be in English was because his instruments were being used not by maths students but by surveyors for measuring and by sailors for navigating – and, unlike mathematicians, neither group could read Latin. Nevertheless, it contained so much new information on logarithms, trigonometry and geometry that one of his contemporaries paid him this tribute: 'He did open men's understandings and made young men in love with that studie [mathematics]. Before, the mathematical sciences were lock't up in the Greeke and Latin tongues and so lay untoucht. After Mr Gunter, these sciences sprang up amain, more and more.'

It was in this book that Gunter first described the chain that was to bear his name: 'for plotting of ground, I hold it fit to use a chaine of foure perches in length, divided into an hundred links'. Four perches measured twenty-two yards, and the fact that this strange distance eventually became integral not only to the game of cricket in his own country (it is the length of the pitch), but to the town planning of almost every major city in the United States (the lengths of most city blocks are multiples of it), was a tribute to the chain's versatility. Its practical advantage was simply that, unlike a rod, its links made it flexible enough to be looped over a person's shoulder, and that being made of metal it neither stretched nor shrank as cords always did. Yet there was more to it than mere practicality. As a passionate believer in the usefulness of maths, Gunter built into his chain the most advanced intellectual learning of the time, until it could almost be compared to a primitive calculating machine.

His cleverness lay in dividing the chain into one hundred links, marked off into groups of ten by brass rings. On the face of it, the chain's dimensions make no sense – each link is a fraction under eight inches long, ten links make slightly less than six feet eight inches, and the full length is sixty-six feet. In fact this is a brilliant synthesis of two otherwise incompatible systems: the traditional English land measurements, which were based on the number four, and the then newly introduced system of decimals, based on the number ten.

It was the Dutch engineer Simon Stevin who first published an account of decimals in 1585, and Gunter was quick to grasp the concept, using them in his logarithmic tables. Where the chain was concerned, he realised that units of ten made for simple calculation, hence the hundred links with the brass rings grouped in tens; but the overall length was no less important. His twenty-two-yard chain measured four rods long, which integrated it into traditional English measurements.

The rod's inconvenient length of 16½ feet was derived from the area of land that could be worked by one person in a day. This was reckoned to be two rods by two rods (thirty-three feet by thirty-three). Thus, there were four square rods in a daywork. Conveniently there were forty dayworks in an acre, the area that could be worked by a team of oxen in a day, and 640 acres in a square mile. All these once variable units became fixed in the sixteenth century, and it was significant that all of them were multiples of four, a number that simplified the calculation of areas.

Gunter's chain produced the happy result that ten square chains measured precisely one acre. Thus, if need be, the entire process of land measurement could be computed in decimals, then converted to acres by dividing the result by ten. With an understandable hint of satisfaction Gunter concluded his description of its use: 'then will the work be more easie in Arithmetick'. It was that ease in calculating acreages, as much as its accuracy and straightforward practicality, that earned Gunter's chain its popularity among surveyors using the old four-based system of measurements. Even the least competent could come close to the standards of exactness that were now expected of them.

It would be difficult to exaggerate the need that the growing army of surveyors had for this kind of assistance. Even an oblong field, where the length was longer than the breadth, made the maths go shaky, and repeatedly the manuals are forced to remind their readers that the area of any square or rectangular field can be calculated by multiplying the length by the breadth. Yet the same mistakes kept recurring: as late as 1688 John Love, who spent many years in Carolina, claimed that he was forced to write his classic work, *Geodaesia*, because 'I have seen so many young men in America so often at a loss ... when a certain number of acres has been given to be laid out five or six times as broad as long.'

The problems multiplied astronomically when the area to be measured was irregular. The surveyors' commonest trick when faced with an irregular shape was to add the lengths of all the sides, divide the total by four, then square the result. The answer thus obtained was quick, easily worked out, and always wrong – but usually not by enough to alarm the land-owner. As one more scrupulous measurer, Edward Worsop, observed of such shortcuts in 1582, 'it is the way all Syrveyors do; – whether it originates in Idleness, inability or want of sufficient pay, it is not for me to determine'. The title of Worsop's volume is self-descriptive: *A discoverie of sundrie errours and faults daily committed by* [surveyors] *ignorant of arithmeticke and geometrie to the damage and prejudice of many of her Majestie's subjects.*

Yet the newly landed gentry of England knew that even an imperfect survey was better than none. While they were measuring and mapping their possessions, then squeezing the highest possible rents from their tenants, the Crown ignored its lands for seventy years after Richard Benese's book came out. In 1603, the Lord Treasurer of England, Robert Cecil, at last commissioned a report on the extent of the Crown's lands, and 'found the King's Mannors and fairest possessions most unsurveyed and uncertain, [their area estimated] rather by report than by measure, not more known than by ancient rents; the estate granted rather by chance than upon knowledge'.

Inefficiently run and casually disposed of, the royal estate which had once produced enough to pay for much of royal government now generated such a small income that the monarch was forced to rely on Parliament to raise taxes in order to run the kingdom. Imperceptibly, power was passing from the land-poor Crown to the land-rich gentry. And a few years later, when it was proposed to plant colonies in the new-found land of America, those who had the money to invest were the same gentry and the merchants, people who could measure their property and count its worth, and not the king.

What Edmund Gunter had devised was a means of making private property. So long as it was the acre that expanded or shrank, while the price remained the same, no true market in land could be established. Once the earth could be measured by a unit that did not vary, supply and demand would determine the price, and it could be treated as a commodity. That was not Gunter's intention, but it was a consequence of the accuracy that was built into his measuring device.

TWO

Precise Confusion

WHAT MAKES MEASUREMENT fundamentally important is that it allows the exchange of goods and services
to take place. Measuring a length of cloth or a herd of animals
or a day's labour gives it a value in terms of size, or number,
or achievement, which enables others in the community to
offer something else – food, protection, even love or loyalty
– of equivalent value. Consequently measurement is almost as
necessary to human society as language, and it occurs in the
earliest civilisations, long before the development of writing.

Incised bone and clay counters used for counting or
exchange have been found at Neolithic village sites in Turkey
dating from the ninth millennium BC – more than ten thousand years ago – while the Sumerian cuneiform script, which
is usually accepted as the earliest form of writing, does not
appear until five thousand years later. It is significant that
many of the clay tablets bearing the cuneiform script refer to
measurements of quantity, weight and length. Without the
ability to measure, co-operative activity could hardly take place.
With it, marketplaces and increasingly sophisticated economies can develop, matching barter, cash or credit to whatever
is owned by one person and desired by another.

What made Edmund Gunter's chain unique among commonly used measurements was that it did not vary. There were
other chains in use, and as late as 1796 an American surveying

manual suggested that in woodland a chain might measure twenty-four yards and in dense forest as much as thirty-two. But wherever a 'Gunter's chain' was specified, it meant precisely one hundred links or twenty-two yards, and an area measuring ten of his chains in length by ten of his chains in breadth would always contain ten acres. In 1607, it was hard to find another measure that was so consistent.

Where modern economics would alter the price of a commodity, the usual practice before the sixteenth century was to change the size of the measure. Confectionery manufacturers do the same today, because children associate a particular price with a particular sweet, and less resentment is caused by shrinking the bar than by increasing the price. For similar reasons, the Vatican city authorities in the fifteenth century required bakers to sell loaves of bread at a fixed price, increasing the size when flour was cheap and reducing it when flour was dear.

Even in the financially sophisticated Netherlands, where banks, corporations and a stock exchange existed from the sixteenth century, it remained the custom in the textiles market to charge the same price per stone (fourteen pounds) for good flax from Zeeland as for the inferior kind from Brabant, but to make up the difference by using a lighter weight for the stone. The Dutch milling trade followed the same pattern – selling flour in larger measures to wholesalers than to retailers, but charging the same to both. At the other end of Europe in the port of Riga, the centre of trade in the Baltic and a major partner in the trading association known as the Hanseatic League, the Latvians quoted a single price on the salt fish they sold to Russian merchants, Ukrainian nobility and their Hanseatic partners, but measured them out in different scales and containers according to the importance of the customer. Accusations about false weights abound in the *cahiers de doléance*, the great lists of complaints that French citizens compiled in 1789, but even they were prepared to accept that using different weights was often fair. As the Sens assembly in Burgundy argued, by using smaller containers to measure out flour in remote villages, traders could charge those far from the market the same price as those nearer to it, and 'the differences in measures are such as to defray the costs of transport'.

Just as an acre of rough pasture was larger than an acre of meadow which could produce hay, so a bushel of oats, which would only make porridge, held more than a bushel of wheat, which could be made into bread, and weak beer suitable only for quenching the thirst was gauged by a large gallon while wine which intoxicated the mind was sparingly measured in a small gallon. The principle applied equally to textiles, wood and other commodities. An ell of coarse cloth was longer than an ell of fine material, a cord of green firewood bulkier than one of dry. Each unit varied not only according to place but according to the benefit it yielded to the buyer. In short, what was being measured was not a quantity but its local, subjective, human value.

The measures themselves were derived from human activity and the shape of the human body itself. Almost universal was the width of the thumb or finger, which was equivalent to an inch, and in the first century AD, Vitruvius gave the classic definition of its relationship to other measures: 'four fingers make one palm, and four palms make one foot; six palms make one cubit; four cubits make once a man's height'. A bushel originated with the amount of seed required to sow an acre of ground – the actual amount varied even more than the size of the acre – while the ell, used for cloth, was either the width of the loom, or the distance from head to wrist (the easiest way to measure woven material is to hold it with an outstretched arm from the chin). Equally organic, and still less exact, were units like the bowshot (the distance an arrow would fly), the *houpée* (how far a shout would carry) and, among the Plains Indians of the United States, the horse-belly view (the furthest a person could see over the prairies when squatting beneath a mustang – approximately two miles).

The weakness of these variable measures was the scope they offered for cheating. From the beginning of writing, and probably earlier, there have been denunciations of those who used false measures. The commonest deceit was simply to use two sets of weights and containers, a large one for buying, a small for selling, and from the earliest times there is evidence of sacred and secular authorities thundering against the practice. 'Thou shalt not have in thy bag diverse weights, a great and a small,' runs the Jewish law in the Book of Deuteronomy, 'but thou shalt have a perfect and just weight, a perfect and just measure shalt thou have, that thy days may be lengthened in the land which the Lord thy God giveth thee.' The Koran inveighs in similar fashion, 'Woe to those who stint the measure. Who when they take by measure from others, exact the full, but diminish when they measure to others, or weigh to them.'

The existence of variable measures, with all their opportunities for fraud, stifled any wider trade because only those actually watching the wheat bushel being filled or the linen being stretched out knew the quantities involved. The success of the great medieval trade fairs at Troyes in Champagne was largely due to the town's capacity to impose its own measures on traders; but it was the Dutch, the first great exporting nation in Europe, who understood the advantage of accuracy better than anyone. According to the English economist Josiah Child, writing in 1668, the booming trade that produced their astonishing prosperity in the seventeenth century was largely based on 'their exact marking of all their Native Commodities'. The Netherlands' major export was fish – herring and cod packed into huge barrels called hogsheads which were supposed to contain sixty-four gallons. Child noted that the Dutch measures were so reliable 'that the repute of their said Commodities abroad continues always good, and the Buyers will accept of them by the marks, without opening; whereas the Fish which our English make in Newfound-Land and New-England, and Herrings at Yarmouth . . . often prove false and deceitfully made, seldom containing the quantity for which the Hogsheads are marked'.

By comparison traders in India, where until the twentieth century every locality had its own, variable measures, found it almost impossible to extend their business beyond the nearest towns. 'I never can tell what I am buying nor how I am selling,' a Madras grain trader complained in 1864. 'My agents inform me that rice is at so much the *seer* [approximately two pounds] [in one village], while in another it is double that price. I take advantage of the opportunity, invest largely, and expect great profits. When the transaction is closed I find I have lost greatly. The *seer* in the first place was perhaps less than half the size of that in the other.' The result was that grain markets in India remained local, and when famine struck in one area, people there died even though food was available elsewhere.

The obvious advantages of exact measuring, and the tragic consequences of local variations, should have made it easy to impose reliable, uniform measures. Yet India has legislation on standard weights and measures dating from four thousand years ago, among the first in the world, while Europe received its inches and pounds from the Romans at the time of Christ. The reason that these systems collapsed is fundamental to the history of weights and measures. However much emperors and rulers might legislate for uniformity, the actual scales and grain containers were held by market traders, landlords and local magnates. They were a source of such profit that no one willingly gave them up, and in every locality throughout history, the clearest guide to where day-to-day power lay has always been the control of weighing and measuring.

It was in 813 that Charlemagne, newly crowned as emperor of most of western Europe by the Pope, issued a famous edict which began '*Volumus ut pondera vel mensurae ubique aequalia*

19

sint et iusta' (We desire that weights and measures should be equal and just everywhere). For nearly a thousand years thereafter, the goal of almost every French king could be expressed euphoniously as '*Un Roi, une foi, un poids*' (One king, one faith, one weight). In 1543, François I asserted bluntly that 'the supreme authority of the King incorporates the right to standardise all measures throughout his kingdom'. But in 1790, according to an authoritative estimate, France possessed thirteen separate lengths for a *pied* or foot, eighteen for the *aune* or ell, and twenty-four for the *boisseau* or bushel. Since the seventy-four parishes of Angoulême near Bordeaux boasted over a hundred different sizes of *boisseau* between them, with one parish alone offering four separate varieties, this was something of an underestimate.

Just as the power of the Dutch trading guilds was demonstrated by their ability to establish uniform measures for exports, so the bewildering variety in France was testimony to the feudal power of the *seigneurs* who retained the right to regulate local weights and measures. The emergence of Gunter's chain as the one measure to determine the dimensions of landed property was due not just to its practicality but to the power of a particular class of people.

On the face of it, England was in much the same position as France. In 960 King Edgar declared that 'the measure of Winchester [England's capital] shall be the standard' for the whole kingdom; but the number of later monarchs who also demanded uniformity – the call for 'one weight and one measure' appears identically in Richard the Lionheart's decree of 1189 and twenty-six years later in Magna Carta – suggests that they were no more effective than their French counterparts.

The most important of these medieval laws, enacted by Henry III in 1266, introduced the sterling system linking weights to coinage, so that there were 240 pennyweights to the pound, a ratio that persisted in the currency for over seven hundred years until 1972 in Britain, and in North America until superseded by the dollar. It failed to prevent most of Henry's successors finding it necessary to pass laws against 'false and deceitful measures', and in 1496 one of them, Henry VII, took the curiously modern step of dumping the sterling pound in favour of a European unit, the Troy pound. Yet in 1588, less than a century later, his granddaughter Elizabeth I had to introduce still more legislation, which she explained was 'called forth by the uncertainty of the weights then in use, to the great slander of the realm and decency of many, both buyers and sellers'.

It is against this background of incessant variation, deceit and falsehood in weights and measures that the precision of Gunter's chain needs to be set. More than any of her predecessors, Elizabeth was responsive to the power of the House of Commons and the people represented there. The contrast between the way she legislated for weight and for length and area was significant.

Where weight was concerned, she found it necessary to add to the Troy system the heavier avoirdupois (meaning literally 'having weight') range, which went from ounces through pounds and stones to hundredweights and tons. Troy was ideal for measuring small items like gold and silver, but the main English export was wool, which was traded in elephantine quantities and in the Flemish markets was always weighed in avoirdupois. It was a concession to variability to use a goldsmith's weights for light objects and a wool merchant's for heavy ones, and a similar surrender appeared in Elizabeth's decision to legislate for a large gallon for measuring beer and a small gallon for wine. Adding to the confusion, sloppy wording of the specifications for containers resulted in four different sizes of bushel being legalised for measuring grains and flour.

By contrast the specifications for length were simple and accurate. A mile was defined for the first time as 1760 yards, instead of the outdated Roman distance of five thousand feet or 1667 yards. This new distance was a deliberate response to surveying needs, for its chief convenience comes in measuring area. It produced a square mile of sixty-four square furlongs, or 640 acres, or 6400 square chains, quantities that a surveyor could easily halve or quarter. In 1601, a brass yardstick of thirty-six inches was constructed as a standard for the country as a whole. Exactness was what the market required, and when Elizabeth's yard was measured in 1797 against the inches, feet and yards used by eighteenth-century scientists, it was found to be precisely 36.015 inches long.

Elizabeth had the energy and administrative skill of a great ruler. She not only ordered new standards to be made for these weights and measures but sent copies to fifty-eight market towns with instructions that a description of them was to be pinned up in every church and read during the service twice a year for the next four years. For good and ill, it is to her that the credit must go for creating a system of weights and measures that was to persist for almost four hundred years, eventually covering all of Britain, and almost a quarter of the globe.

All in all, it was a measuring age. Accurate measurement was becoming vital to the navigation of England's mariners, who used Gunter's cross-staffe and quadrant to find latitude in the trackless ocean, and most notably to Francis Drake in circumnavigating the globe. It was critical to the founding father of the scientific method, Francis Bacon, who advocated measurement and experimentation as the basis of science. When Elizabethans met, they took one another's measure, they danced tightly paced measures like the galliard and volta, and measured their poetry to the short–long rhythm of iambic pentameters. 'Marry, if you would put me to verses or to dance for your sake, Kate, why you undid me,' exclaims rough King Harry wooing his French princess with puns in *Henry V*. 'For the one I have neither words nor measure, and for the other

I have no strength in measure, yet a reasonable measure in strength.'

It was a joke tailored to a particular audience. Without measure, music was noise, poetry babble, and the land wilderness, and none knew it better than the enclosing, acquisitive gentry, the generation whose parents and grandparents first bought their land from Henry VIII, who stamped the Elizabethan age with their energy and imagination, and for whose benefit the legislation on measures was passed.

John Winthrop was just such a man. His family had acquired their five-hundred-acre estate of Groton Manor in East Anglia from Henry VIII, and he himself was a vigorous encloser and improver of the land. It was as much the downturn in rents and farm prices as his Puritan ideals that persuaded Winthrop in 1630 to volunteer to take charge of the colony that the Massachusetts Bay Company proposed to create in Boston. Authoritarian, clear-sighted and charismatic, he was the colony's first governor and imbued it not only with his ideals of communal responsibility and individual conscience, but with his attitude to property.

Although the royal patent gave the colonists the right to settle in New England, there were those, notably Roger Williams, founder of the Rhode Island colony, who felt that the land rightly belonged to the native inhabitants and should first be bought from them. Winthrop summarily disposed of that view with an argument grounded in his own upbringing. 'As for the Natiues in new England,' he wrote, 'they inclose no Land, neither haue any setled habytation, nor any tame Cattle to proue the Land by.'

Since the native Americans had nothing to show that they owned the land, the new Americans could take it freely, and New England, like the Old, would belong to those who could measure it and enclose it. Thus the answer to the question, who owned America? was another question: who would measure America?

THREE

❖

A Hunger for Land

FROM THE ROYAL PALACE at Whitehall, the answer
was simple: the king or the king's representatives would
measure the new-found land. The limits of British America
were defined by map references given in the king's charters,
and the boundaries of its colonies were drawn in the soil by
surveyors appointed by the proprietors and companies to
whom the king had granted the land.

Accordingly, King James I's 1609 charter to the two companies who had put up the money for the Virginia plantation
specified that the London company was to plant its colony 'in
some fit and convenient Place, between four and thirty and
one and forty Degrees of the said Latitude', and the west of
England company based on Plymouth was allocated 'some fit
and convenient Place, between eight and thirty Degrees and
five and forty Degrees'. The four-degree overlap was reduced
in 1620 when a new charter gave the Plymouth company all
the land, to be known as New England, 'from Fourty Degrees
of Northerly Latitude, from the Equnoctiall Line, to Fourty-
eight Degrees of the said Northerly Latitude'. Similar charters
delineated the geographical limits of all the Atlantic colonies
from Nova Scotia to Georgia, often with a final phrase
extending their width 'to the South Sea', in other words to
the Pacific Ocean. A few, like Maryland and Pennsylvania, had
western boundaries fixed in lines or meridians of longitude.

It was the responsibility of the proprietors, until their charters were revoked, to have those degrees of latitude, so easily described in the Privy Chamber in Whitehall, marked out on the ground. The task provoked a sustained wail of complaint from the surveyors who ran the boundaries between the colonies. It was one thing to follow the lie of the land, as the settlers did, zigzagging up from the coast, following rivers and valleys into the foothills of the Blue Ridge mountains or the Alleghenies; it was quite another to run a straight line up the hills, through the swamps and into the unending forest until it emerged into the savannahs of the piedmont. Nevertheless, if the companies and later the royal and aristocratic proprietors named in the charters were to establish their rights of ownership, the boundaries of their colonies and plantations had to be marked westward from the coast.

The most formidable obstacle was the Great Dismal Swamp, a nine-hundred-square-mile expanse of stagnant water, dense bamboo groves and crowded, vine-choked trees lying on the border between Virginia and Carolina. In his account of marking out that border in 1728, *The History of the Dividing Line*, William Byrd II, one of the boundary commissioners, described the surveyors' approach to the swamp: 'The Reeds which grew about 12 feet high, were so thick, & so interlaced with Bamboe-Briars, that our Pioneers were forc't to open a Passage. The Ground, if I may properly call it so, was so Spungy, that the Prints of our Feet were instantly fill'd with Water. But the greatest Grievance was from large Cypresses, which the Wind had blown down and heap'd upon one another. On the Limbs of most of them grew Sharp Snags, Pointing every way like so many Pikes, that requir'd much Pains and Caution to avoid.'

Undeterred, the lead surveyor, William Mayo, pushed through the reeds and disappeared from sight. On the far side of the swamp, Byrd and the other commissioners waited anxiously. After a week, they started to fire off muskets to guide the surveyors, but with no success until on the ninth day the mud-stained party at last emerged, having run the boundary through fifteen miles of swamp.

In his acerbic memoir, Byrd pictures the surveyors as either clowns or heroes. 'Neither the unexpected Distance, nor the Danger of being doubly Starved by Hunger and excessive Cold, could in the least discourage them from going thro' with their Work,' he remarked of the leaders of the survey party, 'tho' at one time they were almost reduced to the hard necessity of cutting up the most useless Person among them, Mr. Savage, in order to Support and save the lives of the rest. But Providence prevented that dreadfull Blow by an unexpected Supply another way, and so the Blind Surveyor escapt.'

The equivalents of Mr Savage were hired to run the line between North and South Carolina in the 1730s after the state split apart. Carolina surveyors, according to John Love, eighteenth-century author of *Geodaesia*, were either corrupt or inept, and the challenge of marking out the boundary, which was to extend from the coast thirty miles south of the Cape Fear river up to the thirty-fifth parallel and then due west along the parallel, defeated the first two parties within a few miles of the coast. Complaining of the 'Extraordinary fatigue [of] Running the said Line most of that time thro' Desart and uninhabited woods', and over rivers and marshland which were breeding grounds for snakes and clouds of vicious mosquitoes, the surveyors refused to return even at the royal salary of £5 a day, five times as much as Mayo was paid for going through the Great Dismal Swamp. Thirty years later, James Cook from North Carolina took on the task but, distracted by 'the rains, the hot weather and the insects' – or so he claimed – ran the boundary eleven miles south of the thirty-fifth parallel and thus took 660 square miles from South Carolina for the benefit of his own state.

Nevertheless, whatever hardships the wilderness threw up, the line had to be run if ownership were to be established. In

1746, Lord Fairfax employed Thomas Lewis and Peter Jefferson to mark off the boundary of his five-million-acre property, virtually a state within a state, that stretched to the Blue Ridge mountains. On 5 October, Lewis wrote of their descent of a mountain in the dark: 'Setting off, we fell into a place that had precipices on either [side], very narrow, full of ledges and brush, and exceedingly rocky. A very great descent. We all like to been killed with repeated falls, and our horses were in a miserable condition. The loose rocks were so [dangerous] as to prove fatal. We at length got to the bottom, not much better, there being a large water course with banks extremely steep that obliged us to cross at places almost [vertical]. After great despair, we at length got to camp about 10 o'clock, hardly anyone without broken [bones] or other misfortune.'

Ten days later the line cut across a swamp: 'Wednesday, October 15th ... The swamp is full of rocks and cavities covered over with a kind of moss [to] considerable depth. The laurel and ivy are so woven together that without cutting it is impossible to force through. In what danger must we be, all places being obscured by a cloak of moss! Such thickets of laurel to struggle through, whose branches are composed of iron! Our horses and ourselves fell into clefts and cavities without seeing the danger before we [fell].'

When they finally got out of the swamp, Lewis wrote in heartfelt relief, 'Never was any poor creatures in such a condition as we! Nor ever was a criminal more glad of having escaped from prison as we were to Get Rid of those Accursed Laurels! From the Beginning of Time, when we entered this swamp, I did not see a [dry] place big enough for a man to lie nor a horse to stand.'

By comparison Charles Mason and Jeremiah Dixon had only the occasional attack by hostile Indians to fear when they were hired in 1763 by the proprietors of Pennsylvania and Maryland to sort out the disputed boundary between the two provinces. There were grades of expertise in surveying, and the equipment provided the best guide. Everyone carried a 16½-foot rod, or Edmund Gunter's invaluable chain, but for professionals a circumferentor, which by the eighteenth century had developed into a theodolite or transit with cross-hairs in the lens of the telescope, and built-in compass and plumb-line, was also necessary. The experts brought along a quadrant or sextant as well for making sun-sights to check their position; in addition to all that, Mason and Dixon had with them a zenith sector built by John Bird, London's foremost instrument-maker. This was a telescope almost six feet long, exactly calibrated and pointing vertically, beneath which they lay flat on their backs to take sightings on particular stars as they passed precisely overhead. Star-charts showing the positions of those stars at different dates and latitudes then enabled them to calculate their latitude with great precision. At the crude end of surveying, anyone able to see straight and multiply and divide could do it, but those at the top end had the scientific exactness and mathematical talent of astronomers.

Taking star-sights with the zenith sector, and sun-sights with the quadrant, cutting a long swathe or 'visto' through the forest for back-sights and fore-sights with the theodolite, and measuring each yard with brass-tipped rods carried in special boxes and calibrated to a five-yard brass standard constructed by the Royal Society of London, Mason and Dixon spent five years on surveying 244 miles at a cost of £3500. To the Calverts of Maryland and the Penns of Pennsylvania, it was worth paying that massive bill to have the extent of their property established beyond doubt. They would have been gratified to learn that the far cheaper line between North and South Carolina was not agreed for another eighty years.

Establishing the exact boundaries of a colony or plantation could be deferred until the population had grown large enough to reach the borders, but from the start every proprietor needed to decide how land inside those boundaries should be measured and settled. There were two models to choose from. In Virginia, the thousand-acre tobacco plantations, and the fifty- or hundred-acre farms granted to each colonist who had paid his own passage, had to be surveyed and registered, but the actual choice of land and of its shape – usually the bottom land along a navigable river with some nearby woodland to provide building material – was left to the landowner.

The early planters developed a crude way of gauging their acreage. Each property was reckoned to run back for a mile from the riverbank. Using the old English measurement of a rod, they simply measured out a length of twenty-five rods along the riverbank, making a straight line which ignored the river's bends. This produced a seemingly awkward distance of 137 yards, one foot and six inches – but multiplied by the 1760-yard depth of the farm, it gave a total of 242,000 square yards, or precisely fifty acres. A hundred-acre farm was fifty rods broad, while a shareholder entitled to five hundred or a thousand acres measured out 250 or five hundred rods. This was frontier maths, and it became second nature to anyone who wanted to own land.

These first farms and plantations were more or less square, but later arrivals fitted in around them, producing crazy patterns of settlement. To define the boundaries of their property they blazed trees or scratched boulders or raised mounds, and described their holdings in terms of these markers. This was the old English practice of using 'metes and bounds' to define the extent of an estate. Thus a surveyor's notes might describe a line as running from the river, 'thence S[outh] 36 [degrees] E[ast] 132 rods to a white oak blazed, thence S 40 W 11 poles to two barren oaks'. Because trees were often destroyed by fire and boulders washed away by floods, boundary disputes filled the courts. It was easier to move on and occupy fresh land far from other claims. 'People live so far apart,' the German immigrant Gottlieb Mittelberger complained in 1756, 'that many have to walk a quarter or a half-hour just to reach their nearest neighbour.'

A different method of surveying evolved in New England, because the climate and soil were harder, and the first colonists arrived as religious groups. It placed the emphasis on communal rather than individual exploitation, and the land was usually granted in rectangular blocks, six or ten miles square, to an association or church which then allocated it to individuals. On 14 May 1636, for example, William Pynchon, Jeheu Burr and half a dozen others were given permission to create a new settlement at Agawam just west of the Connecticut river. The land was to be divided between forty and fifty families, each of which was to have enough property for a house together with some farmland, and parts of a 'hassocky marsh' and nearby woodland. The precise width of each house lot was laid down: 'Northward lys the lott of Thomas Woodford beinge twelve [rods] broade and all the marish before it to ye uplande. Next the lott of Thomas Woodford lys the lott of Thomas Ufford beinge fourteene rod broade and all the marish before it to ye uplande. Next the lott of Thomas Ufford lyes the lott of Henry Smith beinge twenty rod in bredth and all the marish before it, and to run up in the upland on the other side to make up his upland lott ten acres.'

No gaps were left between one individual holding and the next, and one township and the next. The northern settlers might not be able to choose the precise parcel they wanted, but they enjoyed one advantage over the southern planters. In the south, the last remnant of feudalism required landowners to pay the proprietor or colonial government an annual 'quit-rent' of up to two shillings (about fifty cents) an acre, to be quit of the obligations and services they would otherwise owe as vassals. Failure to comply would result in a notice ordering the owner 'to pay your arrears of Quit-Rents and Reliefs and to make your Oath of Fealty' or be fined. In New England, the complication of levying the quit-rent through the church or town soon led to it being abandoned, which meant that freeholders in a New England town effectively owned their farms in fee simple – free of all feudal dues and obligations. They had other social duties – to pay the minister's salary, and attend the church or meeting-house – but their land was undeniably property.

Looking at the two ways of measuring out the land, later proprietors automatically opted for the New England model. The square township, which in New England was known simply as a town, seemed to the aristocratic Carolina proprietors to be 'the chiefe thing that has given New England soe much advantage [in size of population] over Virginia'. They also believed, mistakenly, that this system would give them more control over the colonists. Accordingly their 1665 constitution decreed that all the lowland area, the Tidewater, should be pre-surveyed by a surveyor-general and divided into squares and rectangles 'by lines running East and West, North and South'. From these blocks they proposed to build an American aristocracy, with ordinary immigrants receiving a headright of one hundred acres, and paying quit-rent on them, and above them proprietors, lords of the manor and lesser nobles whose rank depended on the size of their landholding. To ensure compliance, the proprietors instituted a complex system which required the settler to obtain a land warrant from the governor, followed by a survey from the surveyor-general, before the land could be allocated and the claim registered.

Had the colonial proprietors and councils succeeded in maintaining control, the history of North America might have remained colonial. But the idea of property that the colonists carried with them created its own revolutionary current.

The first years after the Pilgrim Fathers landed in Plymouth in the bitter winter of 1620 indicated the direction that history would take. Under the terms of their agreement with their financial backers, they were to work the land in common, sharing the proceeds with the investors in England. The goal of communal land ownership should have been particularly attractive, for they arrived with a close sense of unity arising from the shared desire for religious freedom. Yet in the first years, when they attempted to pool their resources and farm collectively, with young men assigned to work for those who had families, the fields were neglected and they almost starved. In desperation, Governor William Bradford responded to demands that the land be divided up. 'And so,' he noted in his history of the Plymouth colony, 'assigned to every family a parcel of land according to the proportion of their number ... This had very good success for it made all hands very industrious.'

The dramatic increase in yields soon assured the colony's food supplies; but the change came at a cost. 'And no man now thought he could live except he had catle and a great deale of ground to keep them all,' Bradford observed sadly, 'all striving to increase their stocks. By which means they were scatered all over the bay quickly and the towne in which they lived compactly till now was left very thinne.' Religious freedom might have been the settlers' prime reason for sailing to America, but once they were there, the desire to own land came a close second. Or as Richard Winslow put it in his 1624 pamphlet *Good Newes from New England,* 'Religion and profit jump together.'

In 1691 the thinly populated colony was absorbed into the wealthier Massachusetts Bay colony. But it too had changed since John Winthrop had founded it as the shining light of Puritanism, 'the city upon a hill'. By then the Puritan preacher Increase Mather was lamenting that the grandchildren of the original settlers had grown insatiable for land. 'How many men have since coveted after the earth,' he thundered, 'that many hundreds nay thousands of acres have been engrossed by one man, and they that profess themselves Christians have forsaken churches and ordinance, and for land and elbow-room enough in the world?'

In Virginia, the first Jamestown colonists never had that religious sense of cohesion. They only saved themselves from starvation by raiding the farms of Powhatan Indians, and in 1624, eighteen years after the first settlers arrived, it was estimated that massacre and disease had killed six thousand out of 7300 migrants from England. The colony's only source of income came from the sale of tobacco, and that was not enough to prevent the Virginia company from going bankrupt. But what kept the colony alive was a decision in 1618 by one of its shareholders, Sir Edwin Sandys, to attract immigrants by offering a 'headright' of fifty acres of good Virginia soil to anyone who crossed the ocean at his own expense, and as much again for every adult he brought with him. The lure of free land brought a stream of would-be settlers, most of whom died, but by the 1630s the flow of migrants outstripped the death rate from fever, and soon land was being bought and sold at five shillings (about $1.25) for fifty acres.

By the middle of the seventeenth century, in New England and Virginia, land was passing into private hands to be held virtually freehold, except for the quit-rent in the south. So widespread was the process that no one thought it strange, yet for another century this restless hunger to own land made the British colonies unique in North America.

In Mexico and up the Pacific coast, the Spanish acquired land as part of a general pattern of royally sponsored exploration and settlement by the king's representatives. A Spanish civilisation was created in Mexico, with a university, a bishop and a capital housing over fifteen thousand Spaniards, before the first English colonist landed in Massachusetts. The Law of the Indies, enacted in 1573, specified in detail how the colonial government was to lay out towns and settlements. The sites were to be surveyed, religious missions were created to convert the natives, military *presidios* to defend the colonies, and civilian *pueblos* where colonists and colonised could live. It was an empire created from above, belonging to the king and administered as a royal dominion, and even in its final years, during the half-century that Spain ruled California from

1769 to 1821, fewer than thirty individuals were permitted to acquire their own *ranchos* or estates.

For over 150 years, from 1608 when Samuel Champlain established an armed post at Quebec, New France was ruled almost as rigidly. A string of trading ports was established along the Saint Lawrence river, as far as the Great Lakes and down the Mississippi to the Gulf of Mexico. Cities like Montreal and New Orleans were founded, and farms were cultivated in Canada and in the Mississippi delta. Nevertheless, French America was administered feudally. The Crown owned the land and chose who could settle there – Protestants, for example, were banned. It created monopolies to exploit the fur and timber. The *habitant* who actually worked the soil never had clear rights to it. What he owned was the use of the land and the improvements he made to it, but he held the land from a *seigneur* in return for dues and rents, and the *seigneur* held the land from the Crown. French traders and trappers knew the country intimately – they supplied British mapmakers with much of their geographical information – yet by the middle of the eighteenth century barely forty thousand had acquired land outside the main cities.

The land-hunger of the British colonists seemed most bizarre when set alongside the attitude of the native Americans. From the farming Powhatan in Virginia to the Iroquois in New York and the Six Nations in the Appalachians who were primarily hunters, they shared a pervasive understanding that a particular place belonged to a particular people only to the extent that the people belonged to the place. Rights over land were gained only by occupation, long usage or family burial, and these rights were communal, not individual. 'What is this you call property?' Massasoit, a leader of the Wampanaog, asked the Plymouth colonists whom he had befriended in the 1620s. 'It cannot be the earth, for the land is our mother, nourishing all her children, beasts, birds, fish, and all men. The woods, the streams, everything on it belongs to everybody and is for the use of all. How can one man say it belongs only to him?'

Yet the British colonists bought and sold land as though they owned it outright – in fee simple, to use the legal term. Compared to the opportunities offered by New Spain and New France, the Atlantic colonies seemed irresistibly attractive. Little more than a century after the first permanent settlement was established in Virginia, over one and a quarter million settlers were scattered across the wide, empty spaces between the coast and the mountains.

The shape of British America was long and thin, stretching from thirty-one to forty-nine degrees north, a distance of over two thousand miles, but, so far as measured, settled land was concerned, rarely more than two hundred miles deep, a sort of northern Chile. It was in the first years of the eighteenth century that siren voices from beyond the swamps and pine barrens began to tell of the irresistibly fertile ground to be had in the piedmont. 'The best, richest, and most healthy part of your Country is yet to be inhabited,' wrote Francis Makemie in *Plain and Friendly Persuasion* in 1705, 'above the falls of every River, to the Mountains.'

This was the time when the uncontrollable surge of German immigrants, most of them Mennonites and Moravians, followed by the Scotch-Irish, began to move into the area, upsetting Penn's surveyed plan. Pennsylvania alone had an estimated one hundred thousand squatters by 1726, and two-thirds of the colony settled in the 1730s was occupied illegally. Further south in the Virginian piedmont, William Byrd watched crowds of Scotch-Irish squatters taking whatever land suited them, and was reminded of 'the Goths and Vandals of old'. In Massachusetts, settlers moved out into the hilly Berkshires, and in New York up the Mohawk valley, constrained only by fear of French attack. In an attempt to retain control, the Massachusetts government established a string of new townships like Litchfield and Great Barrington in the

Berkshires during the 1720s. Elsewhere, proprietors in Maryland and Pennsylvania, great landowners in New York and northern Virginia, either offered leases to squatters on their land, or tried to drive them out. Royal governors in Virginia and the Carolinas invented schemes to make the squatters legal by offering free land in new townships created on the frontier following the New England model.

None of it worked. The lure of so much property was irresistible. In Virginia, Governor Spottswood himself succumbed to the land rush and claimed eighty-five thousand acres of the newly opened uplands for himself, and in the Carolinas the system of land allocation was overwhelmed by the demand for surveys. Amateur surveyors were hired to help. Wildly unrealistic plats were registered. No one minded. Within two years, warrants were issued for about 600,000 acres, and nineteen thousand of them went to the Governor, while the Assembly Members voted themselves six thousand acres apiece. Between 1731 and 1738 approximately one million acres were registered in Carolina, and when the Surveyor-General, Benjamin Whitaker, complained in 1732 that 'the law enables any common surveyor to perpetuate frauds for his employers through not having to turn his survey into any office', the outraged Assembly sent him to prison for contempt. By then the proprietors had lost all control, and in despair turned the colony over to the Crown.

Georgia's proprietors, particularly the idealistic James Oglethorpe, also intended to survey the colony's territory before distributing it in order to create a slave-free society of smallholders and farmers. In the beautifully proportioned squares and gardens of the capital, Savannah, can be seen all that remains of the plan, for here too the temptation of so much potential property could not be resisted. South Carolinian planters moved across the border, and both they and the Georgians claimed vast estates beyond the Savannah river, outside the squares surveyed by Georgia's founders. In 1751, these proprietors also gave up and returned the colony to royal control.

By that year the population of the Atlantic colonies had risen to over one million, far outstripping that of New France and outnumbering the Spanish-descended inhabitants of New Spain, who had been there for more than two hundred years. Families were larger than in the Old World because farms were bigger and more hands were needed to work them, but there was also the lure of unclaimed acres that drew the dispossessed from the far side of the Atlantic. If they could not pay their own passage, they came as indentured servants, willing to act as near-slaves for a number of years for the chance of eventually acquiring property.

Even in Elizabeth I's reign, the enclosures in England deprived so many villeins and labourers of the common land and common grazing on which they depended to keep their families alive that they were forced to beg in the towns, giving rise to a series of ferocious laws against 'sturdy rogues and vagabonds' and 'wandering beggars'. They were joined in the seventeenth century by the Diggers, radical Puritans who had fought for Oliver Cromwell in the English Civil War and who, having defeated the King, tried to reclaim common land by digging and cultivating it – hence their name. 'True religion and undefiled is to let everyone quietly have erth to manure [cultivate],' wrote their leader, Gerrard Winstanley, 'that they may live in freedom by their labours.' But Cromwell and his generals were property-owners to a man, and promptly turned them out. When enclosure reached Scotland in the eighteenth century, and improving landlords in the Lowlands and clan chiefs in the Highlands took as their own the land that their tenants and clansmen once held in common, the newly dispossessed provided more raw material for the colonies.

There was a certain irony that these newcomers should now be amongst the hungriest of all America's property-owners, relying on the surveys and chains that had driven them off their homeland. But in 1628 the landed gentry in Parliament had forced King Charles I to accept the Petition of Right which guaranteed the right of the property-owner not 'to be put out of his land or tenements ... without being brought to answer by the due process of law'; and none knew better than the dispossessed how powerful was the lure of owning a farm which could not be taken away.

It was from England that the idea of land as property had originally come, and it was no coincidence that with it had arrived Gunter's chain – twenty-two precisely calibrated yards, each exactly thirty-six inches long (plus that miserly 0.015 of an inch extra that would not be discovered for almost two centuries) – and the practice of showing an estate's exact extent on a surveyor's plat drawn to scale. From England came too the philosophical underpinning developed by John Locke that the individual earned a right to property by 'mixing his labour' with what had once been held in common. Every landowner who had ever enclosed, manured and improved a field understood this proposition perfectly, but by 1750 the American idea of property had evolved further.

What surveyors like George Washington, Peter Jefferson and Daniel Boone were doing was speculating on the future value of land. However much they could earn from surveying fees, it was dwarfed by the profits to be made from buying good land cheap. 'The greatest Estates we have in this Colony,' the young George Washington acknowledged after a summer spent surveying the vast Fairfax estates, 'were made ... by taking up & purchasing at very low rates the rich back Lands which were thought nothing of in those days, but are now the most valuable Lands we possess.' In 1752, at the age of twenty, Washington purchased 1459 acres in Frederick County, in the Virginian piedmont, the first step in a career of land-dealing that eventually made him owner of over fifty-two thousand acres spread across six different states. He usually 'improved' his holdings by clearing them of trees, but for most speculators their property rights did not depend on any idea of 'mixing their labour' with the soil. Their sole claim to ownership lay in the survey and the map that came from it.

In 1751 Benjamin Franklin stated openly what was apparent to authorities on either side of the Atlantic, that the population of the colonies was growing at such speed it would double to 2.6 million by 1775. It would not be long, he predicted provocatively, before 'the greatest number of Englishmen will be on this side of the water'. To some Americans, that prospect raised constitutional questions about being controlled from across the Atlantic; but to the planters of Virginia and the Carolinas, and to financiers in New York and Philadelphia, it also indicated that the purchase of American land was a wise investment. Nowhere was it cheaper than west of the range of mountains known generally as the Appalachians, but divided from south to north into the Blue Ridge, the Alleghenies and the Adirondacks.

In 1756, a South Carolina surveyor, John William de Brahm, was sent to build a fort at Loudon on the Little Tennessee river on the other side of the mountains, in country that still belonged to the Cherokee Indians. 'Their vallies are of the richest soil, equal to manure itself, impossible in appearance ever to wear out,' he reported back in admiration. 'Should this country once come into the hands of the Europeans, they may with propriety call it the American Canaan, for it will fully answer their industry and all methods of European culture, and do as well for European produce ... This country seems longing for the hands of industry to receive its hidden treasures, which nature has been collecting and toiling since the beginning ready to deliver them up.'

Control of all this desirable territory as far west as the Mississippi still remained with the French but in 1763 they were forced to cede it to the British following their defeat in the French and Indian War. Soon other surveyors took the chance to follow de Brahm. Their findings were brought together in a famous map by Thomas Hutchins, not published until 1778, but whose attractions were known a decade earlier.

On the map's crackling parchment, the Appalachians appear as a black, impenetrable mass of cross-hatching running from the bottom left-hand corner to the top right-hand corner; but west of them are broad rivers and rolling hills denoted by lines which curl gently towards the Mississippi and are interspersed with Hutchins' own observations in neat italic writing: 'A rich and level country', 'Very large natural meadows; innumerable herds of Buffaloe, Elk, Deer, etc feed here', and along the Wabash river, 'Here the country is level, rich and well timber'd and abounds in very extensive meadows and savannahs; and innumerable herds of Buffaloe, Elk, Deer, etc. It yields Rye, Hemp, Pea Vine, Wild Indigo, Red & White clover etc.'

Not even the promise of dancing-girls could have inflamed the colonial appetite more than the prospect of such fertility. That the land belonged to the Cherokee, Shawnee and Six Nations was a detail that could be overcome by personal negotiation, as Judge Richard Henderson and surveyor Boone did, or by killing and terror, as numerous others did. To the planters it was obvious that, with the French claims removed, the entire area between the mountains and the river now lay open for occupation.

But ownership of land is never simple. It includes rights not just to the soil, but to the metals below, the vegetation above, the sunlight and the air; to its use, development, access and enjoyment; and to much more that, for a fee, a lawyer will reveal. Since any or all of these may be rented, leased, loaned or distributed in different ways, landed property is usually described as a bundle of legal rights which can be split up and dealt with separately. Although no one can now claim all of them outright – environmental laws and national needs limit the owner's rights – under feudal tenure they all belonged to the King. Thus, much of the 1629 charter creating the Massachusetts Bay Company is made up of lists of different types of land, forms of ownership and the way they are to be transferred. King Charles I promises to 'give, graunt, bargaine, sell, alien, enfeoffe, allot, assigne and confirme' to the Company all the 'Landes and Groundes, Place and Places, Soyles, Woods and Wood Groundes, Havens, Portes, Rivers, Waters, Mynes, Mineralls, Jurisdiccons, Rightes, Royalties, Liberties, Freedomes, Immunities, Priviledges, Franchises, Preheminences, Hereditaments, and Comodities'. Nevertheless, ultimately the land still remained the King's, to be held by the Company 'in free and common soccage' – which meant that having given, granted and all the rest, the Crown retained an overriding, feudal competence over that part of America.

It was because they were part of that feudal structure that the original proprietors had charged a quit-rent in place of feudal dues. But most of the proprietors had gone now, defeated by the settlers' uncontrollable desire for land, and the colonial legislatures, such as Virginia's House of Burgesses, were effectively forums for the colonists' interests. It was easy for settlers, squatters and speculators looking longingly towards the land beyond the Appalachians to forget the King's feudal power.

Then, on 7 October 1763, came a harsh reminder of the legal reality behind American property. By royal proclamation, George III declared it 'to be our Royal Will and Pleasure ... that no Governor or Commander in Chief in any of our Colonies or Plantations in America do presume for the present, and until our further Pleasure be Known, to grant Warrants of Survey, or pass Patents for any Lands beyond the Heads or Sources of any of the Rivers which fall into the Atlantic Ocean from the West and North West'. In effect, a line had been drawn along the watershed of the Appalachians beyond which land could not be measured and owned, and everyone who had already settled west of it was commanded 'forthwith [to] remove themselves from such settlements'.

The King had the right to order this, because legally all the land in British America was his; but it planted feudal authority full in the path of the property-seekers.

FOUR

Life, Liberty and What?

T HERE WERE MANY STRANDS leading to the moment when the colonists felt driven to weave their anger together into a single declaration of opposition to rule from London. The decision of the British Parliament to close the port of Boston in 1774 as punishment for the destruction of a valuable cargo of tea brought to the surface the resentment of northern merchants already burdened by duties on their goods, a general fury at the earlier killing of civilian rioters by British troops, and a pervasive fear that colonial assemblies were powerless against the King's ministers. But that autumn, when delegates of the discontented colonists convened in Philadelphia as members of the First Continental Congress in order to articulate their grievances, it was not by chance that the first resolution they agreed was 'That they are entitled to life, liberty, & property . . .'.

Here property meant more than land alone, but for Virginians especially it was land as property that they had in mind, and in particular land beyond the Appalachians. Hence the declaration of the first paragraph of the Virginia constitution, drawn up in June 1776 by George Mason, 'That all men are by nature equally free and independent, and have certain inherent rights . . . namely, the enjoyment of life and liberty, with the means of acquiring and possessing property, and pursuing and obtaining happiness and safety.' Entitlement to this sort of property was a subject on which the humblest Conestoga mule-driver was at one with the grandest planter.

Barely ten years earlier, the reaction of Colonel George Washington to the royal veto on the acquisition of land beyond the mountains could have served as a warning of what was to come. There had been no more loyal and energetic commander in the French and Indian War, but the Colonel was also a Virginian planter and land speculator, and his views were widely shared.

'I can never look upon the Proclamation in any other light (but this I say between ourselves) than as a temporary expedient to quiet the minds of the Indians,' Washington wrote in 1767 to his colleague and fellow-surveyor Colonel William Crawford. 'It must fall, of course, in a few years, especially when those Indians consent to our occupying those lands. Any person who neglects hunting out good lands, and in some measure marking and distinguishing them for his own in order to keep others from settling them, will never regain it.'

George Washington was not a man to be deflected even by his sovereign's express command, although as a serving officer he deemed it best to be discreet. 'If you will be at the trouble of seeking out the lands,' he continued to Crawford, 'I will take upon me the part of securing them, as soon as there is a possibility of doing it and will, moreover, be at all the cost and charges of surveying and patenting the same ... By this time it will be easy for you to discover that my plan is to secure a good deal of land. You will consequently come in for a handsome quantity.'

Alas, poor Crawford never did. The King's veto was still nominally in force when he was taken prisoner by a Cherokee band while leading a column of troops in territory beyond the Appalachians. On 3 August 1782 the *Virginia Gazette* carried a report of his ordeal by a Dr Knight, who had been captured along with Washington's colleague: '... the unfortunate Colonel was led by a long rope to a stake, to which he was tied, and a quantity of red-hot coals laid around, on which he was obliged to walk bare-footed, the Indians at the same time torturing him with squibs of powder and burning sticks for two hours, when he begged of Simon Gurry (a white renegade who was present) to shoot him. [Gurry's] reply was, "Don't you see I have no gun." [Crawford] was soon after scalped and struck several times on the bare skull with sticks, till being exhausted, he laid down on the burning embers, when the squaws put shovel-fuls of coals on his body, which made him move and creep till he expired. The Doctor was obliged to stand by and see this cruelty performed; they struck him in the face with the Colonel's scalp, saying "This is your great Captain's scalp, tomorrow we will serve you so." '

Gruesome stories such as these were used to justify acts of equal cruelty on the other side. John Heckewelder, a Moravian missionary living with the Tuscarawas Indians, told in 1773 of the Indian-hunters 'who maintained that to kill an Indian was the same as killing a bear or a buffalo and would fire on Indians that came across them by the way – nay more, would decoy such as lived across the river to come over for the purpose of joining them in hilarity; and when these complied they fell on them and murdered them'.

These atrocities were evidence of the mounting conflict between land-hungry colonists and native inhabitants. Consequently when George III banned land purchases beyond the mountains, it was, as the proclamation worded it, so that 'the several Nations or Tribes of Indians with whom We are connected, and who live under our Protection, should not be molested or disturbed'. But even Americans who might have sympathised with this strategy could not accept the King's feudal right to impose the ban.

The power that the land beyond the mountains exerted on people's minds can be deduced from the attempts to circumvent the veto. Two land companies had been created to speculate in the west before the ban was in place – the Ohio

Company, which proposed to purchase 500,000 acres beyond the Ohio river, and the Loyal Land Company, organised by Peter Jefferson with investment coming mostly from his neighbours in Goochland County, Virginia, which aimed publicly at buying 800,000 acres of Kentucky, but privately had ambitions of exploring and acquiring millions more as far west as the Pacific Ocean. In the fifteen years after George III's proclamation, a whole succession of similar speculative ventures came into being.

The Mississippi Company was created in 1768 to settle land along the river with George Washington as one of its founders, followed by the Illinois and Wabash in which Patrick Henry had an interest, and the Watauga Association which settled eastern Tennessee and later tried to establish the independent state of Franklin. In 1775 Judge Richard Henderson of North Carolina sent Daniel Boone, a brave scout but an incompetent surveyor, to find territory in southern Kentucky, where he signed a treaty with the Cherokee Indians giving Henderson's Transylvania Company several million acres – Boone's surveying lapses made it unclear exactly how much land was involved. The most ambitious of them all, the Vandalia Company, which aimed to acquire sixty-three million acres in what is now Illinois and Indiana, employed Benjamin Franklin as its London agent and even counted among its members such influential figures in the British government as the future Prime Minister, Lord North, and the Lord Chancellor, Lord Camden. 'One half of England is now land mad,' remarked one of its promoters, 'and everybody there has their eyes fixed on this country.'

These powerful interests created some loopholes in the prohibition, but it was still in place in 1773 when a young surveyor named Rufus Putnam sailed up the Mississippi to Natchez. There he began surveying over a million acres on the banks of the river so that it could be sold to New England veterans of the French and Indian War. He was a rarity among the mostly southern land speculators, because he came from Massachusetts. His career is instructive because it illustrates how widely the effects of the ban were felt. In character he more or less resembled the coat of arms he later adopted which showed three bristly wild boars below a roaring lion surrounded by thistles and the motto in spiky Gothic lettering, 'By the name of Putnam', and almost everything he achieved he owed to his ferocious determination.

In 1745, when Rufus was barely seven years old, his father died, leaving the family destitute. His mother's second marriage was to an illiterate drunkard named Sadler. 'During the six years I lived with Capt Sadler,' Rufus wrote bitterly, 'I never Saw the inside of a School house except about three weeks.' At the age of fifteen he apprenticed himself to a man who built watermills, and sucked up knowledge where he could; but as he confessed in his autobiography, 'having no guid I knew not where to begin nor what corse to pursue – hence neglected Spelling and gramer when young [and] have Suffered much through life on that account'.

Rufus's prospects were transformed by the French and Indian War. In 1757 he joined the Royal American Regiment, where he became an engineer, a trade that taught him how to carry out every kind of measurement. When peace came in 1763 the orphan used his new skill first to build mills, and later to survey land. The demand for surveyors in British America was such that a good practitioner could command an income that matched a lawyer's. Even at the age of seventeen, George Washington was able to boast of his earnings to a friend. 'A doubloon is my constant gain every day that will permit my going out,' he wrote, 'and sometimes six pistoles.' Since a doubloon was worth about £15, and six pistoles around $22.50, a good week might bring in around $100; even as President he hardly earned much more.

Soon Rufus felt secure enough to marry the wealthy Persis Rice, daughter of Zebulon, who in the first six years of their marriage bore four daughters and a son. What he wanted, he wanted immediately. Neither in private nor public life did he ever show any guile, and rarely much patience. With five children to feed, and the expectation of more on the way, he turned to land speculation, surveying and acquiring land in the West Indies and what is now Alabama for New England veterans. The Natchez venture was on a far larger scale, and like George Washington, who was trying to secure land on the Ohio river for Virginia veterans of the same war, Rufus Putnam proposed to keep some of the best sites for himself.

It was Richard Henderson who put into words the dream that drew Rufus and thousands more into the western territory beyond the Blue Ridge and the Alleghenies. 'The country might invite a prince from his palace merely for the pleasure of contemplating its beauty and excellence,' he wrote in praise of his company's Kentucky acres, 'but only add the rapturous idea of property and what allurements can the world offer for the loss of so glorious a prospect?

When that rapturous idea was snatched away by the proclamation of a feudal monarch three thousand miles away, it was small wonder that the colonists should have felt the need to state not simply their inalienable right to life and liberty, but also to the acquisition of property. Typically, when Rufus too fell victim to the royal prohibition, he preferred action to words. He had spent eight months laying out the ground for future settlements at Natchez when the Board of Trade in London issued an order specifically forbidding any further surveys on the Mississippi. Returning home to Massachusetts, he found his personal frustration echoed in the accumulating anger of the colonists.

Persis had kept the family going during Rufus's long absence, and nine months after his return to the marriage bed, another son was born; but by then he was soldiering again. This time he was fighting the King, one of the first men to be commissioned into the Massachusetts Regiment following the first ringing shots at Lexington in April 1775.

Almost at once, his talent for finding practical solutions to complex problems found an outlet. Following the outbreak of hostilities, an American force of 11,500 men under George Washington surrounded a British army under General Thomas Gage in Boston. With little faith in his untrained men against six thousand British regulars, Washington needed to pen Gage in, but lacked anyone with the knowledge to construct the necessary siegeworks. In the same uncomplicated way that he had taught himself to be a surveyor, Rufus read up a book on military engineering, then laid out a system of trenches and defensive posts that kept the British cooped up until the spring of 1776, when the threat of artillery bombardment forced them to sail away. Washington rewarded his bristly subordinate by appointing him chief engineer to the army, and to the end of his days Rufus remained Washington's man, body and soul. In peacetime, all political questions were solved by doing whatever Washington required, and if that was not clear, Rufus would write to ask. Anyone who opposed his general and later his president was an enemy – and one particularly prominent opponent, whose name Rufus could never bring himself to mention, was an 'Arch Enemy'.

That individual was Thomas Jefferson. The time for him to reveal himself as Rufus Putnam's Arch Enemy did not come until after the war, but even in 1776 it was evident that he marched to the beat of a different drum. Indeed, if there was any one person immune to the general lust for land beyond the Appalachians, it was this Virginian planter, who in his wording of the Americans' Declaration of Independence changed their fundamental assertion of rights from 'life, liberty and property' to 'life, liberty and the pursuit of happiness'.

It is one of the greatest paradoxes in the paradoxical character of Thomas Jefferson that he – who was to acquire more western land on behalf of the United States than any speculator could have dreamed of possessing, who laid the foundations for the nation's further territorial expansion to the Pacific by sending Lewis and Clark to find a route to the coast in 1803, who believed passionately in the virtues of owning land, and adored his own plantations and garden at Monticello – was so tepid in acquiring it for himself. From his father, Peter, he inherited about seven thousand acres in the Virginian piedmont including Monticello and its farms, and his marriage to Martha Wayles Skelton in 1772 added to his holding. He even joined several schemes for acquiring land beyond the mountains, but then neglected them, and all eventually failed.

By birth and upbringing, Jefferson belonged to the Virginian plantation aristocracy – the family had been there since the late seventeenth century – and generations of Jeffersons had followed the frontier west; to them land acquisition was second nature. In other matters he had many of the characteristics of the planter class, especially their extravagance: living constantly in debt, but insisting on the best in wine, furniture, carriages and harness, he died a bankrupt in all but name. As a child, Jefferson was clearly destined for that aristocratic role. The model before him was that of his large, raw-boned father, Peter, who had not only run the Fairfax line with Lewis through the swamps and ravines of the Blue Ridge, but later extended William Byrd's boundary between Virginia and Carolina almost into Kentucky, and only returned home to build a plantation mansion on the frontier and lay plans with his neighbours to acquire still more land through the Loyal Land Company. Before his death in 1757, when Thomas was fourteen, Peter appointed various friends and relatives as guardians to his children, and all but two of these were trustees of the Loyal Company.

Thomas's first years were spent roaming free in the foothills of the Blue Ridge. Recognition of his place as an insider came with his election in 1769 as a twenty-six-year-old lawyer to the Virginia legislature's House of Burgesses, which represented plantation owners' interests. When he began to question the basis of the royal claim to exercise feudal power over the land beyond the mountains, his conclusion was what his class would have expected – that George III had no right to restrict their desire to acquire property.

In Saxon England, before William the Conqueror had imposed his regime, Jefferson argued, feudalism was unknown. 'Our Saxon ancestors held their lands, as they did their personal property, in absolute dominion,' he declared in a fiery pamphlet, *A Summary View of the Rights of British America*, published in 1774. It was William and his Norman invaders who had invented 'the fictitious principle that all lands belong originally to the king', and the fiction had been maintained by his successors. Since America had been occupied and won without help from the Crown, George III had no grounds for claiming power over the disposal of its land. Only a democratically elected legislature had that power, Jefferson concluded, and, in a phrase that would have been music to any settler's ears, wrote that if it failed to do so 'each individual of the society may appropriate to himself such lands as he finds vacant, and occupancy will give him title'. It was this sweeping attack, cutting at the very foundation of royal power over America, that led to Jefferson's appointment by the Continental Congress in 1776 to the three-man committee responsible for drafting the Declaration of Independence.

Yet Jefferson was never a truly typical member of his class. He thought about land in a way that no speculator would. To the end of his life he retained an idealised vision of pre-feudal Saxon society, with its local court and administration based on the 'hundred' or parish, and its values derived from the stout-hearted, independent-minded yeomen farmers who worked the soil. Consequently all the political systems he devised, for counties as for nations, shared one fundamental quality: the widest possible distribution of land.

In a memorable passage in *Notes on the State of Virginia*, the book written from 1780 to 1782 which expresses some of Jefferson's deepest beliefs, he explained, 'Those who labour in the earth are the chosen people of God, if ever he had a chosen people, whose breasts he has made his peculiar deposit for substantial and genuine virtue.' Other trades had to depend 'on the casualties and caprice of customers. Dependence begets subservience and venality, suffocates the germ of virtue, and prepares fit tools for the designs of ambition.' But '[c]orruption of morals in the mass of cultivators is a phaenomenon of which no age nor nation has furnished an example'.

It is hard to think of any other Virginian who might have entertained such a far-fetched idea. Not William Byrd, who observed of his fellow-planters, 'Our land produces all the fine things of Paradise, except innocence.' Not Washington, cheated by a farming acquaintance, William Clifton, into paying an extra £100 (about $470) in 1760 for an estate neighbouring on Mount Vernon. Not Jefferson's friend Fielding Lewis, who remarked of the land deals in the piedmont that 'every man now trys to ruen his neighbour'. For them land was a source of wealth, not the basis of God-given virtue.

There was, however, more wisdom in the world than the planters knew of.

In 1760, at the age of sixteen, Thomas Jefferson had been sent to William and Mary College in Williamsburg, where he had fallen under the influence of a remarkable teacher named William Small, and the experience marked him for life. In his autobiography, written in his seventies, Jefferson paid Small this tribute: 'It was my great good fortune, and what probably fixed the destinies of my life that Dr. Wm. Small of Scotland was then professor of Mathematics, a man profound in most of the useful branches of science, with a happy talent of communication, correct and gentlemanly manners, & an enlarged & liberal mind.' Jefferson was famously guarded in his emotions – this prompted Joseph P. Ellis, an especially perceptive observer, to call his biography of Jefferson *American Sphinx* – but where Small was concerned he became almost fulsome. 'Dr. Small was . . . to me as a father,' Jefferson confided to a friend. 'To his enlightened and affectionate guidance of my studies while at college, I am indebted for everything.' The brief reference in his autobiography to his real father, the great surveyor, is stark by comparison: 'My father's education had been quite neglected; but being of a strong mind, sound judgment and eager after information, he read much and improved himself.'

In 1760 Willliam Small was just twenty-four, a product of those Scots universities which bred in the likes of David Hume and Adam Smith a restless desire to find a rational key to understanding human nature and human society, and to six-teen-year-old Thomas Jefferson he must have seemed like an ideal elder brother. 'He, most happily for me, became soon attached to me & made me his daily companion when not engaged in the school,' Jefferson remembered gratefully, 'and from his conversation I got my first views of the expansion of science & of the system of things in which we are placed.'

The most obvious consequence of Small's company was that all his life Jefferson preferred to think of himself as a scientist rather than a politician. 'Nature intended me for the tranquil pursuits of science, by rendering them my supreme delight,' he wrote soon after his retirement from the presidency, 'but the enormities of the times in which I have lived have forced me to take a part in resisting them and to commit myself on the boisterous ocean of political passion.' Politics, even sometimes the United States itself, he would refer to as 'an experiment'. In his house he hung portraits of Francis Bacon, modern science's founding father, and Isaac Newton, its great-est luminary. He studied botany, the most highly developed science of the day, and prided himself on his membership of the American Philosophical Society, the country's leading scientific body. In old age, he boasted to his friend and enemy

John Adams, 'I have given up newspapers in exchange for Tacitus and Thucydides, for Newton and Euclid, and I find myself much happier.'

It was Isaac Newton above all who had expanded science and shown how the system of things was made. The consequences of Newton's laws of motion were fundamental to the development of physics and astronomy in general. In Newtonian physics there were theoretical explanations for every event, from the movement of the planets to the swing of a pendulum, and two brilliant analyses from his monumental *Philosophiae Naturalis Principia Mathematica* (Mathematical Principles of Natural Philosophy), published in 1687, provided what would prove to be the real starting-point for modern measurement. Newton deduced that instead of being a perfect sphere, the earth bulged at the equator and flattened at the poles; consequently, gravity would be stronger at the poles, because they were closer to the centre of the earth. Once the shape of the earth was known, the possibility of estimating its size offered eighteenth-century scientists a foundation for the first new way of measuring things to be devised in ten thousand years.

But Newton's laws also taught a lesson that went beyond physics. In the universities of Aberdeen, Edinburgh and Glasgow, the belief was encouraged that just as it was possible to discover through rational enquiry the fundamental principles that governed the natural world, so reason made it possible to understand the principles governing human nature. This, the essential outlook of the movement known as the Enlightenment, was what William Small passed on, and what Jefferson meant by understanding 'the system of things in which we are placed'.

'Fix reason firmly in her seat, and call to her tribunal every fact, every opinion,' he would later advise his young nephew, in an echo of Small's teaching. 'Question with boldness even the existence of a god; because, if there be one, he must more approve of the homage of reason, than that of blindfolded fear.' Once rational enquiry had uncovered the principles of human nature, it would be possible to establish the conditions under which individuals could be allowed as much liberty as they desired, and yet a rational, self-regulating society would emerge.

Owning land, for example, would give society's members an interest in building a law-abiding, democratically based community, and education would teach them how best to use their freedom. The ideal, therefore, was to distribute land and schools as widely as possible. By the same criterion, the accumulation of too much land by one person had to be opposed, because it prevented others acquiring it. Nothing illustrates better the difference between Peter and Thomas Jefferson than the father's restless search for more acres, and the son's unrelenting opposition to land speculators whose profits depended on driving up the price of empty land. 'Whenever there are in any country uncultivated lands and unemployed poor,' Thomas wrote soon after independence had been won, 'it is clear that the laws of property have been so far extended as to violate natural right.'

That was the wisdom that Jefferson derived from William Small and his circle of friends. It appealed profoundly to a young man who, as Joseph Ellis observed, craved for 'a world in which all behavior was voluntary and therefore coercion was unnecessary, where independence and equality never collided, where the sources of all authority were invisible because they had already been internalised'.

In 1764, Small returned to Britain, but Jefferson never ceased to feel gratitude to the man who, he acknowledged, 'filled up the measure of his goodness to me, by procuring for me, from his most intimate friend G[eorge] Wythe, a reception as a student of law under his direction, and introduced me to the acquaintance and familiar table of Governor Fauquier, the ablest man who had ever filled that office. With him, and at his table, Dr. Small & Mr. Wythe . . . & myself, formed a *partie quarrée*, & to the habitual conversations on these occasions I owed much instruction.'

After independence was declared in July 1776, Jefferson returned to Virginia, where his reputation as the Declaration's prime author led to his appointment to a committee revising the state's laws so that they reflected republican rather than royalist values. It is a telling fact that at this, his first opportunity to put theory into practice, he put forward what might be called the full Small programme.

The best-known of his proposals is the 'Statute for Establishing Religious Freedom', a landmark in legislative tolerance; but no less significant were otherwise banal bills to change inheritance law so that a landed estate could be left equally to all the children rather than to the eldest son alone, and to prohibit clauses in a will preventing later generations dividing up land among several heirs. At the same time, Jefferson introduced a bill to give seventy-five acres to any Virginian who did not already have any land, and to offer a 'head-right' grant of fifty acres to every landless immigrant who arrived in the state from overseas. Together with Virginia's generous promise of land to soldiers enlisting in its regiments – ranging from a hundred acres for enlisted men up to fifteen thousand acres for a major-general – Jefferson's land grant proposals would have created a network of small farms guaranteeing the future health of democracy in the state.

To complete the Saxon ideal, Jefferson came up with a plan for wholesale administrative and educational reform – or as he put it, 'I drew a bill for our legislature, which proposed to lay off every county into hundreds or townships of 5. or 6. miles square, in the centre of each of which was to be a free English school.'

As any of his fellow Virginians could have warned him, Virginia politics did not work like that. When the state came to dispose of its unoccupied land in 1783, the Act that finally emerged from horse-trading in the Assembly was designed to sell the territory with as little restriction as possible, and the process turned into a swill-bucket for speculators. Surveyors were bribed into setting aside the best plots, land warrants were acquired cheaply from army veterans, and wads of the state's devalued paper currency, which carried the right to claim unoccupied land, were bought up for a quarter of their value.

One speculator, Robert Morris, acquired one and a half million acres of western Virginia, while another, Alexander Walcott, secured a million acres. On top of these claims, the Virginia legislature allowed speculators to benefit from any inaccuracies in the survey up to 5 per cent of the total – thereby adding a free bonus of seventy-five thousand acres to Morris's allocation – and then generously added a clause permitting purchasers to keep anything more than that which might inadvertently have been included as a result of the 'ignorance, negligence or fraud of the surveyors'.

As a dry run for Jefferson's far more spectacular experiment involving the territory west of the Appalachians, this was a humiliating failure. Yet it was less wounding than a weakness of temperament that was exposed by the War of Independence – in moments of crisis, it became clear, emotion was liable to overcome Jefferson. In 1779 he was elected Governor of Virginia, but when the British army moved into the state in 1780, instead of calling out the militia he evidently froze, leaving the decision to be taken by others and allowing the state capital, Richmond, to be overrun. The inquiry launched into Jefferson's conduct he described as 'a wound on my spirit which will only be cured by the all-healing grave'. But evidence that this was not a freak reaction came soon afterwards in a second, more personal crisis.

In the autumn of 1782, Jefferson's wife Martha died at the age of thirty-three after giving birth to their sixth child. They had been married for less than eleven years, and in that time Martha had been almost constantly pregnant, with six live children born, although only two survived beyond infancy, and three other pregnancies which ended in miscarriages. At her death, Jefferson was prostrated by a grief so consuming that for a month he could not face anyone and stayed secluded in a room where, his daughter recalled, he wept and groaned, emerging at last only to go for long, solitary horseback rides in the mountains. There may have been an element of guilt in this – the risk of repeated pregnancies to the health of delicate women was well understood – but whatever the source of his emotions it was obvious that the force of them made it impossible for Jefferson to function. In retreat from the pain of grief, he threw himself into work which absorbed all his energies and attention.

One year earlier, in October 1781, Lord Cornwallis and his British army, hemmed in by Washington on land and by the French fleet at sea, had surrendered, and American negotiators were now in Paris deciding the terms of the peace. At home the Continental Congress, which represented the nearest thing to a central government that the thirteen states could agree on, was attempting to work out the new nation's future government and how to pay off the mountain of debt accumulated in paying for Washington's continental army. Its single asset, if the negotiators could prise it from Britain's grasp and the states were prepared to give up their own claims to it, was the land between the Appalachians and the Mississippi.

'There are at present many great objects before Congress,' wrote the Rhode Island delegate, David Howell, early in 1784, 'but none of more importance or which engage my attention more than that of the Western Country.'

It might be sold to pay the country's debts, it might be divided up to create new states, it might be administered on a new model, it might be made over to the pre-revolutionary land companies. The course decided upon would help to determine the relationship of the central government to the states.

In June 1783 Jefferson was elected a delegate from Virginia to the Continental Congress, and immersed himself in the many great objects before it, above all in the question of the western lands. Preserved in the Library of Congress are pages of his comments on proposed legislation, and draft bills whose margins are filled with his detailed annotations. In the space of less than a year, from June 1783 to May 1784, this escape into mental work produced numerous contributions to United States law, and three measures so substantial that they were to permeate every aspect of American life: the invention of the dollar, the procedure for creating new states from the Western Territory, and the means of surveying that territory. Had Jefferson had his way, there would have been a fourth: the invention of a new set of weights and measures. It indicates the cohesion of his thinking that all four formed part of a single logical structure.

FIVE

Simple Arithmetic

G ENERAL RUFUS PUTNAM had had a good war in every sense. His rank was a reward for the sterling service he had rendered Washington both at the siege of Boston and during the fighting in New York. When the continental army's strategic retreat in 1776 pulled the focus of the war further south, Rufus had returned home to command the 5th Massachusetts Regiment in defence of his own state. There he had found time to father three more children, two in the dark days of defeat, and a third to celebrate the approach of victory – not for nothing were there rampant boars on the Putnam coat of arms – and to acquire the fine estate of Rutland that formerly belonged to a wealthy loyalist. But he was not content to rest on his laurels. With the return of peace, he was impatient to stretch his surveyor's chain across the wide-open spaces beyond the Ohio river.

In April 1783, Timothy Pickering, a delegate to the Continental Congress, reported that 'there is a plan for the forming of a new State Westward of the Ohio. Some of the principal officers of the Army are heartily engaged in it. The propositions respecting it are in the hands of General Huntington and General Putnam, the total exclusion of slavery from the State to form an essential and irrevocable part of the Constitution.'

Rufus was not a speculator – his stand against slavery, effectively ruling out land sales to Southern planters, was evidence of that – but he was proposing to acquire as much land as any of the pre-revolutionary companies, almost eighteen million acres from the Ohio to Lake Erie. Nevertheless, he wanted to divide this land up into '756 townships of six miles square', because, as he told his former commander George Washington, 'I am much opposed to the monopoly of lands and wish to guard against large patents being granted to individuals . . . it throws too much power in the hands of a few.' Instead, he hoped to see the entire area settled by veterans of the continental army who could acquire land either by using the warrants issued on completion of service or by paying a fixed, small price. This would have the double benefit of settling a part of the United States vulnerable to British invasion from Canada with reliable soldiers, and of reviving the value of the military warrants, whose worth was 'no more than 3/6 & 4/- on the pound [i.e. less than 20 per cent of the face value] [but] which in all probability might double if not more, the moment it was known that Government would receive them for lands in the Ohio Country'. Rufus petitioned Congress to grant him the land, and when it did not respond he wrote again to George Washington in April 1784 to ask, as was his habit, what should be done.

'The Settlement of the Ohio Country, Sir, ingrosses many of my thoughts and much of my time since I left Camp,' he wrote, and the delay was making the veterans impatient. 'Many of them are unable to lie long on their oars waiting the decition [sic] of Congress on our petition.'

Washington, who had retired from command of his victorious troops and returned to his estates, was no less impatient for Congress to reach some decision about the land beyond the mountains. Throughout the war a stream of settlers had moved into Tennessee and Kentucky, and the increase in their numbers after the fighting was over prompted Washington to warn the Congress that without some policy, 'the settling, or rather overspreading of the Western Country will take place by a parcel of *Banditti* who will bid defiance to all Authority'. The thorny pre-revolutionary conflict between settlers and squatters, proprietors and Goths, had not gone away.

Yet nothing could be done until the states agreed to give up individual claims to territory that they had all won from the British. Virginia, for example, as the original colony, had some rights to all the land from Lake Erie west to modern-day Wisconsin and south to St Louis, while Massachusetts could point to a phrase in its charter giving it 'the mayne Landes from the Atlantick ... on the East Parte, to the South Sea [the Pacific] on the West parte'. States like Maryland and New Jersey whose western boundaries had been drawn by surveyors refused even to sign the Articles of Confederation, which bound them to act together against the British, until these gigantic claims had been abandoned.

Although Rufus Putnam seems not to have been aware of it, the key to the deadlock was in the hands of the Arch Enemy. In 1781 Jefferson as Governor had ceded Virginia's claim to the Western Territory to the Continental Congress. One by one, the other claimant states followed suit, and the Articles of Confederation were at last signed in 1781, shortly before the war ended. But true to his Enlightenment self, Jefferson had added a reservation. Only the United States government could acquire the territory from the native American nations who owned it. Consequently any claims made by pre-revolutionary land companies were cancelled. Congress, however, was filled with company sympathisers. No United States territory could exist until one or other side backed down.

Over the next twenty years Jefferson was to engage in an ideological war with land speculators whose interests were diametrically opposed to his. In the Continental Congress their ringleader was Robert Morris. He had out-manoeuvred Jefferson over Virginia's disposal of land within her existing boundaries, and was now the Congress's Superintendent of Finance, an influential position which helped ensure that the congressional mood remained in favour of the land companies.

The two men were polar opposites: Morris, whose fat, friendly, asthmatic appearance distracted attention from a cold, abacus mind, and the lean, controlled, complex Jefferson, concealing his emotional weakness and high-flown idealism behind a stream of words and studied informality. 'His whole figure has a loose, [shambling] air,' observed Senator William Maclay of Jefferson in 1790. 'He has a rambling vacant look, and nothing of that firm, collected deportment which I expected . . . He spoke almost without ceasing. But even his discourse partook of personal demeanour. It was loose and rambling, and yet he scattered information, wherever he went, and some even brilliant sentiments sparkled from him.'

Unlike Jefferson's privileged background, Morris's past was one of unremitting effort from his arrival as a penniless immigrant from England in 1747, through long years as an accountant working for the wealthy Philadelphia merchant Charles Willing, until he was made a partner in Willing's company, and became one of the wealthiest men in America. During the war he had used his wealth to underwrite contracts for the purchase of supplies and munitions for Washington's army, and with the goodwill this earned he secured still more profitable contracts for himself.

The second skirmish in Jefferson and Morris's long campaign occurred over currency.

When George Washington replied to Rufus Putnam in April 1784, his letter illustrated the basic money problem facing the new republic. Pointing out that Congress was still deadlocked on the land question, Washington offered instead to lease his thirty thousand acres in the Ohio valley to the impatient Massachusetts veterans. The rental would be high, about $36 per hundred acres, he explained, because 'it is land of the first quality' and the cost of improvements he had made amounted to '£1568 Virginia, equal to £1961/3/3d Maryland,

Pennsylvania or Jersey currency'. If Rufus was still not sure how much that meant in Massachusetts, Washington added that 'a Spanish milled dollar shall pass in payment for six shillings'.

The handicap to Washington's real-estate deal was one that hobbled every commercial transaction in the United States at that time. Although the legal tender remained officially the British pound, divided into twenty shillings, each in turn subdivided into twelve pennies, its value in America differed from one state to the next. The commonest single coin, the Spanish dollar, was worth five shillings in Georgia, but thirty-two shillings and sixpence across the border in South Carolina, and six shillings in New Hampshire, while the official London rate was four shillings and sixpence. Still more confusingly, it was divided into eight bits in Pennsylvania, but contained ten bits in Virginia. Along with Spanish dollars and doubloons, there were also French louis d'or and écus; Portuguese moidores, pistoles and half-Joes, so called because they carried the image of King Johannes V; Dutch florins; Swedish dollars or riksdalers; as well as the sovereigns, shillings and pennies of Britain. Familiarity taught most people to juggle all these currencies, and just as the teenage George Washington casually reckoned up his pay in pistoles and doubloons, so Thomas Jefferson, scribbling a quick note of a sale of land, recorded that the price had been '200 [pounds] of which 20 half-Joes are paid', or $950 and $160 respectively. Opinion in Congress, however, held that a single currency was needed to help hold the new nation together.

The first recommendation came from Congress's Superintendent of Finance, Robert Morris. It was based on an unrealistically small unit, a fraction of a penny, and in the opening shot of their campaign, Jefferson replied with a report early in 1784 recommending instead the adoption of the Spanish dollar as the most convenient basis of the new currency. In the interests of simplicity he suggested that instead of being divided up into eight bits, it should be decimalised.

'Every one remembers,' he told Congress, 'that when learning money arithmetic, he used to be puzzled with adding the pence, taking out the twelves and carrying them on; adding the shillings, taking out the twenties and carrying them on. But when he came to the pounds where he had only tens to carry forward, it was easy and free from error. The bulk of mankind are school boys thro' life. These little perplexities are always great to them.' Accordingly, the dollar should be subdivided into tenths or dismes, hundredths or cents, and thousandths or mills.

It was an argument that everyone could understand, and less than eighteen months later, on 6 July 1785, Congress resolved that 'the money unit of the United States of America be one dollar', and that 'the several pieces shall increase in decimal ratio'. This was not just an intellectual victory for Jefferson; it effectively prevented Morris from achieving his goal of running the United States Mint, a source of potentially enormous profits.

In the course of the currency debate, Morris had declared, 'it is happy for us to have throughout the Union the same Ideas of a Mile and an Inch, a Hogshead and a Quart, a Pound and an Ounce'. Even without their earlier hostilities, this would have set him on a collision course with Virginia's representative, for it was clear to Jefferson that the rationale for replacing pennies and shillings with a decimal unit applied equally to American weights and measures.

Officially each state had adopted the system of Troy and avoirdupois that Elizabeth I had imposed on sixteenth-century England and that subsequent legislation in London had amended; but barely a single unit was the same from one state to the next – except for Gunter's chain and the acre. A Virginia tobacco-grower like Thomas Jefferson measured his crop in hogsheads, well aware that a Virginia hogshead was larger than a New York hogshead but smaller than one from Maryland, and that a tobacco hogshead from any state was a different size to a brewer's hogshead. A Boston brewer might also refer to his hogshead of beer as a pipe, butt or puncheon, knowing that each of them contained two cooms, four kilderkins, eight rundlets, or sixty-four gallons. But a Baltimore brewer who used the same measures somehow ended up with only sixty-three gallons of beer in his Maryland hogshead, while the number of gallons in a Pennsylvania brewer's hogshead actually changed depending on where the beer was sold, because the law required innkeepers to sell beer inside the inn by the wine gallon, which was smaller than the beer gallon that was used for selling beer outside the inn. And the confusion over liquid measurements was nothing compared to the labyrinth of quarts, gallons and bushels used for measuring corn or flour. Because of flaws in English legislation, each of them could be one of eight different sizes, and might be measured either heaped above the brim of the container, or struck, meaning level with the brim, as custom or the local market dictated.

Round three of the Jefferson–Morris war was, therefore, bound to occur over weights and measures. The direction of Jefferson's ideas can be found on a sheet of paper dating from the spring of 1784 and headed innocuously 'Some Thoughts on a Coinage', which shows that he conceived of the dollar and a new, decimal American set of weights and measures as being two parts of a single system. The weights would be co-ordinated with the dollar, so that a pound would equal the weight of ten dollar-coins. The lengths were to be derived from the size of the earth.

This idea, first mooted by Abbé Gabriel Mouton in 1665, had been given scientific respectability by the work of Jean Picard, the father of modern geodesy or earth measurement. The circle of the equator is divided into 360 degrees, and each degree is subdivided into sixty minutes. The distance of one of those minutes was equal to one nautical mile, a unit that navigators had used since the sixteenth century, and which remains in use today by pilots, mariners and other navigators. Picard's successor, Jacques Cassini, had estimated the total distance round the equator to be more than twenty-five thousand miles (today's best figure puts it at 24,902 miles, or 40,075 kilometres), which made each degree a little less than seventy statute miles.

Notes and tables soon fill the page, to be followed on a separate line by Jefferson's calculation for the length of a minute, or 'geographical mile', in his words: 'Then a geographical mile will be of 6086.4 feet.' Acknowledging the difficulty of physically measuring the equator, he comes up with a way of checking the length of this new mile: 'A pendulum vibrating seconds is by S[i]r I[saac] Newton [calculated to measure] 39.2 inches'.

It was Galileo, allegedly dreaming in church and watching the slow swing of a chandelier, who first noted that the amount of time a pendulum took to move through its arc from one end to the other depended on its length. The longer the pendulum, the more time it needed – to be exact, the time was proportional to the square root of its length. In London, Isaac Newton's calculations had shown that the swing of a pendulum 39.1682 inches long took exactly one second (nearer the pole, the stronger pull of gravity would quicken the pendulum fractionally), and it was this scientifically testable unit – known as a second's pendulum – that Jefferson proposed to use to check the length of his mile.

By the time he starts to compare his new decimal lengths with traditional units, the geographical mile has already become the American mile in his mind:

> Then the American mile = 6086.4 [feet].
> English = 5280 f[eet].
> furlong = 608.64 f[eet]. = 660 [feet]
> chain = 60.864 f[eet]. = 66 [feet]
> pace = 6.0864 [feet] fathom = 6 [feet]

The widest discrepancy was with the English mile, but perhaps to comfort himself Jefferson lists all the other miles in use, from the Russian – barely fifteen hundred old yards – and ascending through the Irish, Polish and Swedish to the Hungarian, which stretched for almost seven old miles. In such company there would be nothing strange about the American mile. There the 'Thoughts' end, a remarkable race through what was evidently a vast fund of knowledge stored in Jefferson's mind.

It was no academic exercise. Jefferson intended to apply his new system to the most important subject facing the new republic – the measurement of the Western Territory.

That same spring, the Continental Congress, desperate to raise money from the sale of its land, had at last accepted the Virginian condition, and on 1 March 1784 Jefferson led his state's delegation formally ceding its claims to the immense region to the north-west of the river Ohio. For the first time the United States had a territorial reality to match the spiritual identity outlined in the Declaration of Independence.

On the same day, and as part of the deal, a committee chaired by Jefferson produced a report on how the Western Territory was to be governed. It covered each stage of the process, starting with the land's acquisition from the Indians – it could be obtained only by the United States government – through the delineation of boundaries, choice of government, and eventual admission as states to the United States. Once the land had been acquired and surveyed, settlement could take place, and as the region filled up with people, they could apply for their territory to become one of the states of the Union on a level of equality with the original thirteen founders.

Even the names of some of the proposed states were specified, among them Michigania and Illinoia, which more or less survived, and Assenisipia and Metropotamia, which did not. What was striking was their shape. Except for river and lake boundaries, all were defined by parallels of latitude running east–west, and meridians of longitude running north–south. The Atlantic seaboard states which had given up claims to the territory also had their western borders chopped off straight on a meridian. Consequently the future shape of the United States would not be long and thin, but square and geometrical.

The next day Jefferson was appointed chairman of a committee to choose the best way of surveying and selling off the land inside those imaginary states, and when it reported on 30 April 1784, it too opted for squares.

The Virginian method of allowing purchasers to choose their property and of surveying it by metes and bounds was ruled out. Instead it was to be surveyed before occupation and marked out in squares aligned with each other, following the New England model, so that no land would be left vacant. At Jefferson's insistence these squares were to be called 'hundreds', while Hugh Williamson, another committee member, was probably responsible for proposing that their sides should run due east and west, and north and south. But the square was integral to what the geographer W.D. Pattison termed 'their ambitious attempt to realise a dream of democratic rationality for the American West'.

To those who had not read Jefferson's 'Thoughts on a Coinage', the report's second sentence detailing the dimensions of those squares must have appeared inexplicable: '[The Western Territory] shall be divided into Hundreds of ten geographical miles square, each mile containing 6086 feet and four tenths of a foot, by lines to be run and marked due north and south, and others crossing these at right angles.' The hundreds could be sold entire or divided into lots measuring one geographical mile square. Whether or not his fellow legislators understood Jefferson's thinking, they would certainly have recognised the political wisdom of tacking a potentially unpopular measure onto one that was both desirable and vitally important. The United States would have the opportunity not only to raise money by the sale of land, but as a bonus would have the first decimalised system of measurements in the world.

In a letter written a few days later to an old friend, Francis Hopkinson, Jefferson gloated over the audacity of his plan. 'In the scheme for disposing of the soil an happy opportunity occurs for introducing into general use the geometrical mile in such a manner as that it cannot possibly fail of forcing it's [sic] way on the people,' he began. There would be objections, he acknowledged. Legislators would argue that the report could not be passed into law because it bore 'some relation to astronomy and to science in general, which certainly have nothing to do with legislation'. He imagined crusty conservatives advocating the retention of both the penny and the inch in order to 'preserve an athletic strength of calculation'. But, he predicted, all opposition was bound to fail: 'This is surely an age of innovation, and America the focus of it!'

In those first years, the republic had substance, but was not yet formed, and the quality that marked its leaders was the certainty that all their actions helped give it shape. The second President, John Adams, thought of his role in terms of making a watch. 'When I consider . . . that I may have been instrumental in stretching some Springs and turning some Wheels,' he wrote to his wife, 'I feel an Awe upon my Mind which is not easily described.'

Jefferson saw himself as the architect of a self-regulating, land-owning democracy. *Notes on the State of Virginia* contained his goal for the disposal of the western lands: 'the proportion which the aggregate of the other classes of citizens bears in any state to that of its husbandmen [farmers],' he stated, 'is the proportion of its unsound to its healthy parts, and is a good-enough barometer whereby to measure its degree of corruption. While we have land to labour then, let us never wish to see our citizens occupied at a work-bench, or twirling a distaff.' In the short term, the health of the republic might depend on raising cash, but in the long term, 'It is the manners and spirit of a people which preserve a republic in vigour.' The future of the United States depended on settling as many of its citizens on the land as possible. Democracy would grow from decimals, squares and surveys.

The land plan in its entirety was Jefferson's, and only he could act as its advocate. But in the summer of 1784, before he could present his report, he was appointed the United States' envoy to France. In his absence, a new chairman was appointed to the committee. Changes were made, and the decimal measures were the first to go. What Congress took from Jefferson's report was the grid pattern, with the east–west lines cutting the north–south at right angles, and the land survey before sale. The dimensions of the squares, however, owed more to Edmund Gunter than to Thomas Jefferson.

In the Ordinance which Congress passed on 20 May 1785, for 'disposing of lands in the western territory', it was laid down that 'the surveyors shall proceed to divide the said territory into townships of 6 miles square, by lines running due north and south, and others crossing these at right angles, as near as may be . . . The lines shall be measured with a chain; shall be plainly marked by chaps on the trees, and exactly described on a plat.'

The thirty-six-square-mile townships were to be divided into one-square-mile lots, four of which in each township were reserved to the government 'for the maintenance of public schools'. Every alternate township was to be sold whole, and the intervening ones by square-mile lots. The surveyors were also to make note of prominent features like salt-licks, mines, mills, mountains and the quality of soil, and their compasses were to be adjusted to due north.

Rufus Putnam was delighted by the proposals, which were almost identical to those he advocated, but in a report to the absent Jefferson, James Monroe commented discreetly, 'It deviates I believe essentially from your [recommendations].' Nevertheless, the measuring of America could at last begin.

SIX

A Line in the Wilderness

THE POINT OF THE BEGINNING had been decided by a boundary commission headed by two of the United States' finest surveyors, Andrew Ellicott, who would later help to lay out the plan for the nation's new capital on the Potomac, and David Rittenhouse, whom Jefferson declared to be the greatest astronomer in the world. Their task was to mark out the western boundary of Pennsylvania, running it north until it cut the Ohio river and then on towards Lake Erie. That boundary had been specified in the original charter to William Penn. Until Virginia ceded her claims to the Western Territory, everything beyond that limit had theoretically been hers – all of Kentucky, most of present-day Ohio, Indiana and as far west as Wisconsin. Once the line had been drawn, it would become part of the Northwestern Territory that Thomas Hutchins was to survey.

The calibre of the boundary commissioners was shown by their use of a zenith sector that Rittenhouse and Ellicott had designed and constructed, as well as a theodolite and quadrants. There was never going to be any doubt about the accuracy of the western boundary of Pennsylvania. On 20 August the party reached the river, and one of the surveyors, Andrew Porter, recorded: 'This morning continued the Vista over the hill on the south side of the River and set a stake on it by the signals, about two miles in front of the Instrument [the theodolite], brought the Instrument forward and fixed it on a high post, opened the Vista [i.e. cut a line through the trees] down to the River, and set a stake on the flat, on the North side of the River.' It was this stake that marked the starting point of the land survey.

On a September afternoon exactly 215 years later, with the dogwood and oak turning in the crisp sunshine, it was tempting to imagine the scene before Hutchins; but the present Ohio woodland is largely made up of secondary growth, mere matchstick trees with trunks barely two feet across in place of the eight-foot-diameter giants burned by the settlers to clear the land for farming. The colours seen by Hutchins were carried by a range of native species that had grown largely undisturbed for centuries. 'The whole of the above distance is shaded with black and white Walnut Trees,' he wrote, 'also with Black, Red, and an abundance of White Oaks, some Cherry Tree, Elm, Hoop-Ash and great quantities of Hickory, Sarsarfrax, Dogwood, and innumerable and uncommonly large Grape Vines, producing well tasted Grapes of which Wine may be made.'

From a surveyor's point of view, such open forest was preferable to scrub, where every yard had to be hacked clear so that the chainmen could measure out the land. Nevertheless, of the forty-strong team that camped with Hutchins beside the Ohio river, the largest group consisted of heavy-shouldered axemen recruited to cut down trees that might obscure the view. Since the land belonged to the United States, each of the thirteen states was supposed to have sent a surveyor to take part in the measurement, but only eight had bothered to make the two-week journey by coach from Philadelphia to Pittsburgh over tracks pitted by potholes and tree-stumps, and then by horseback to the river. Significantly, at least three of them were advance men for land speculators.

Thomas Hutchins must have seemed the ideal choice as leader. He was a good enough surveyor to be part of the

Pennsylvania boundary team, and his maps were prized by soldiers and by other cartographers. In his military career, senior officers had repeatedly recorded his sense of initiative, and as an explorer he had seen more of the continent than any other English-speaker alive. Everywhere he had shown an abundance of that indispensable frontier quality, self-reliance. But he was the wrong man in the wrong job.

Temperamentally Hutchins was a loner, drawn to the unexplored wasteland by its emptiness. Even his obituarist could not help remarking on his 'unconquerable diffidence and modesty', which was so marked that when Hutchins was orphaned as a teenager he could not bring himself to ask relatives for help. Instead he joined the army, where anonymity came easily. Showing an aptitude for numbers, he secured promotion first as paymaster then as an engineer preparing plans and fortifications for his unit, the Royal American Regiment, during the French and Indian War. Unlike more sociable colleagues, he devoted his time to maps and notes, and in that thoughtful vein contributed an appendix in 1765 to a regimental war memoir which has become more famous than the main work.

In it he outlined a model for the construction of frontier settlements, recommending that they should be made up of one-mile-square parcels of land arranged around a square defensive hub on a riverbank. Some historians have given Hutchins credit for thus creating the blueprint for the land survey, but his claim is no stronger than Putnam's, or of those who advocated keeping things simple for dumb surveyors. Squares solved many contemporary problems, but long after the last settler was scalped and the last surveyor learned trigonometry, it was the ideas embodied in the squares that continued to make them powerful. And Jefferson planted the most potent of those.

As well as this interest in right angles, the two Thomases shared an enthusiasm for the potential of the west. Writing up his observations of Louisiana, where he was stationed after the French and Indian War, Hutchins described a land of plenty, where German plantation-owners grew 'indigo, cotton, rice, beans, myrtle-wax and lumber', and the French 'employed themselves in making pitch, tar, and turpentine, and raising stock, for which the country is very favourable'. As in the north, it was the woodland that was the country's chief glory. 'The quantity of lumber sent from the Mississippi to the West India islands is prodigious,' Hutchins observed, 'and it generally goes to a good market.' He presented a vision of riches not simply for information but deliberately to encourage his fellow countrymen to move into this rich territory. 'If we want it,' he concluded stirringly, 'I warrant it will soon be ours.'

It was while Hutchins was in London overseeing the printing of maps that the revolution broke out and the defining moment of his life occurred. He still held a commission in what had been the British army, and senior officers not only expected him to return to active service, but offered him promotion. Hutchins held out against the inducement, steadfastly refusing to fight against his own people. At length he was arrested in London, but preferred to lose his maps, money and commission rather than change his loyalties. When he was released, he found his way back to the United States with a letter from Benjamin Franklin testifying to his unshakeable nature, and George Washington promptly appointed him geographer to his southern army.

In over forty years of adult life, integrity had been the hallmark of everything that Thomas Hutchins did. But over the next two years, as leader of the surveying expedition, he displayed a degree of ineptitude that could only have come from the incompatibility of his temperament with the task he had been given.

Using a sextant to check his position against the sun and the Pole Star, Hutchins established that the survey's starting point lay on the latitude of 40 degrees 38 minutes and 02 seconds north. Modern methods indicate that he was actually twenty-five seconds, or about 850 yards, further north than he calculated, but since he was not using a zenith sector, it was an acceptable error. From that point the surveyors were to proceed due west, climbing up from the river, working through the choppy hills and valleys of the Allegheny plateau, until they reached those open, level meadows Hutchins' own maps had described as lying beyond the mountains. With a compass, Hutchins took bearings on a distant mark, then sent the axemen forward to clear a path for the chainmen.

The task of the chainmen was described in Robert Gibson's 1739 *Treatise of Practical Surveying*. The rear man stood by the starting stake with one end of the chain, while the front man, carrying the other end and a set of tally pegs, walked towards the mark unrolling the chain as he went. The *Treatise* stressed the importance of the rear man insuring that the front one was always in line with the mark. 'If the hinder chainman causes the foreman to cover the object,' Gibson wrote, 'it is plain the foremost is then in a right line towards it. The inaccuracies of most surveys arise from bad chaining, that is from straying out of the right line.'

At the end of twenty-two yards, a tally peg was inserted, the rear chainman came up, and the process was repeated. Ten chains made a furlong, eighty chains made a mile, 480 chains made one side of a township. At each mile, they put in a marker post. The axemen who accompanied the chainmen chopped away trees and bushes, leaving a long trail behind them. As they moved forward, the surveyor checked the position by taking compass bearings on a tall tree or an exposed hilltop, then blazed the tree or marked the hill, and entered the details in his notebook. So they moved through the virgin forest like a caterpillar, hunching up and stretching out, drawing a straight line to the west.

In fact the land was far from virgin. It had been worked and occupied for centuries by peoples like the Delaware, Ottawa, Shawnee and Miami, who alternated between summers farming in the forest and winters buffalo-hunting in open country further west. Pressure from white settlers attempting to cross the Ohio, and from the Iroquois who claimed much of the territory, had pushed a dozen tribes from as far away as Wisconsin to form themselves into a loose alliance known as the Western Confederacy.

The contrast between the surveyors' straight line and the Indians' winding trails and inconspicuous farm plots could not have been more stark. 'We do not understand measuring out the lands,' a Shawnee spokesman later declared at an abortive peace conference. 'It is all ours. Brothers you seem to grow proud because you have overthrown the king of England.' The Delaware and the Miami, however, who lived closest to the Ohio river, did understand that colonial attitudes to ownership were radically different from their own, and sensed the danger in the surveyors' straight line. It was not they who had given permission for the survey, but the Iroquois, who in 1784 signed a treaty at Fort Stanwix, ceding to the United States the territory they claimed west of the Ohio. Consequently the two peoples watched the progress of the survey with growing anger.

Hutchins' party were aware of the hostility, and proceeded carefully, until they received word that an American trading post in a Delaware village some miles away had been raided. The news, as Hutchins heard it, was that the trader had been killed, and 'all the signs of war were left behind, [the Indians] having marked the inside of the door and contiguous trees with red Paint'.

For Hutchins' team, it was warning enough. On 8 October, barely a week after starting, they pulled out, and retreated all the way back to Pittsburgh. They had surveyed just four miles. Hutchins' report made much of the dangers and difficulties they had encountered, and particularly the problem of estab-

lishing true, as opposed to magnetic, north due to local variations in the magnetic field. Nevertheless, it was an inglorious
beginning, and the team, which was being paid $2 a mile,
could not have been happy. Congress sent Hutchins back the
next year with instructions to run his line westward for only
seven ranges – a distance of just forty-two miles – and, disregarding the land to the north of it, which had been allocated
to Connecticut, to concentrate on surveying southwards to a
bend in the Ohio river.

The following year the party was strengthened by more surveyors, including notable figures like Israel Ludlow and
Winthrop Sargent, both of whom were acting for the Ohio
Company of Associates, a land company that intended to buy
in the area. With a military guard provided by General Josiah
Harmar's troops to keep the Indians at bay, the work went on
more quickly. Every six miles, a new survey party started running a line due south, and a fresh east–west line was begun
six miles below the Geographer's Line, as Hutchins' base-line
was known. Yet by the winter only half the area that Congress
had expected to have available for sale had been surveyed.

The work was not completed until June 1787, and it was
not only late but shoddy. The Geographer's Line did not run
due west, but drooped about two degrees to the south. Few
of the east–west lines crossed the north–south ones at right
angles, and as a result the intended grid of squares had
become a grid of diamonds and irregular quadrilaterals. 'The
surveyors were apparently all individuals,' C. Albert White
commented dryly in his magisterial work *A History of the
Rectangular Survey System*, 'with individual concepts of how
to comply with the "six miles square" and "right-angle"
requirements.'

There had to be a reason for work so lamentably below his
usual standards, and one was clearly that poor, solitary Hutchins could not establish authority over his crew. But there may
have been another. Before his chainmen had taken their first

step, the government's team faced competition from two private land companies, the Ohio Company and an individual speculator, John Cleves Symmes. At least three of Hutchins' surveyors, Ludlow, Sargent and Absalom Martin, were agents for his competitors, and it was not in their interests that the government should be selling land before they were ready.

Neither his character nor career had equipped Hutchins to deal with worldly businessmen – as one of his obituaries put it, 'all join in declaring him to have been "an Israelite indeed, in whom there is no guile" ' – and he may never have realised that his efforts were being sabotaged. But it is also possible that eventually he too sold out.

The territory covered by the Seven Ranges, as it came to be known, was never going to be good land. George Washington, who knew it well, and owned thirty thousand acres on the other side of the Ohio, thought the only workable soil was at the bottom of the valleys, while the rest was useless except 'to support the Bottoms with Timber and Wood'. Yet Hutchins' 1787 report suggested it would be ideal for arable crops of most kinds. 'The Land is too rich to produce Wheat,' he admitted, 'but it is well adapted for Indian Corn, Tobacco, Hemp, Flax, Oats, etc.' Perhaps this was simply part of his programme to encourage Americans to move west, but that same year the Ohio Company had just bought over one million acres next to the Seven Ranges, and it was now in their interest as well as the government's to talk up the values of the North-western Territory. When the Company's sales pamphlet appeared, with lavish descriptions of the agricultural, mineral and timber riches in the area, it boasted a presumably well-paid endorsement from Thomas Hutchins, Geographer of the United States: '*I do Certify* that the facts therein related are judicious, just and true, and correspond with observations made by me during my residence of upward of ten years in that country.'

If Hutchins had allowed himself to be pressured by commercial interests and then to be bought by them, he was not the sort of man who could live with the consequences. In 1787 he left the Seven Ranges survey to join a team establishing the boundary between Massachusetts and New York; but he was already sick. He returned for a few weeks to the Ohio, and in the spring of 1789 quietly died at the age of fifty-eight. An affectionate obituary in the *Gazette of the United States* paid tribute to his upright character, ending with the poignant phrase, 'He has measured much earth, but a small space now contains him.'

What Thomas Hutchins left behind was the rough prototype of a system that would eventually cover much of the country he loved so strongly. For over two centuries every sale of real estate within that area has been recorded on maps which stand in direct line of descent from the plats he first drew. It constitutes an impressive memorial to a good man.

The public lands started to be sold at auction in New York City at the end of September 1787. Fewer than 100,000 acres were actually paid for, and only $117,108 reached the United States Treasury. Disappointed by the returns, the Continental Congress suspended the public survey, strengthening the voice of those who wanted to sell the territory simply to raise money rather than to serve as an experiment in democracy. It was argued that the grid itself could never work, because it ignored geography. Some squares might have no water supply while others consisted of nothing but swamp, and George Washington himself warned that 'the lands are of so versatile a nature, that to the end of time they will not by those who are acquainted therewith be purchased, either in Townships or in square miles'.

From a Jeffersonian point of view, the sales carried a worrying message, for despite all precautions more than half the land that was sold went not to individuals but to a couple of speculators, Alexander Macomb and William Edgar, associates of his *bête noire*, Robert Morris. Nevertheless, the unique advantage of the grid system was demonstrated when the very first patent was issued at the New York City land office on 4 March 1788. It went to one John Martin, who paid $640 for a one-square-mile section of the Northwestern Territory, identified as Lot 20, Township 7, Range 4. Even now this is an almost anonymous parcel of field, wood and rock in Belmont County, Ohio, and then it was simply part of the forest; but once it had been surveyed and entered on the grid, it could be picked out from every other square mile of territory, and be bought from an office three hundred miles away on the coast. Whether the gain was enough to outweigh the grid's disadvantages, however, was a matter still to be decided.

The first to benefit from the decision of Congress to sell direct to the land companies was the Ohio Company of Associates, whose originator and guiding spirit was the redoubtable Rufus Putnam. The Company was to make him rich, but that was not his original intention. It started as the realisation of the idea he had outlined to Washington, to settle New England veterans in the strategically important Western Territory. The main features of his plan were strikingly Jeffersonian. The territory was to be divided into 'townships of six miles square'; following the model of Putnam's native Massachusetts, land would be set aside for schools, and slavery would be made illegal.

Soon after the Ohio Company was set up in January 1786, the nature of this uncommercial plan changed. Among its eight hundred shareholders were indeed many soldiers from New England, but they also included speculators like Robert Morris, and much of the Company's capital of $250,000 was used to purchase military land warrants and depreciated currency on the open market for as little as ten cents on the dollar. None of this financial subtlety sounds like Rufus's work. What makes him attractive is the quality that Persis must have loved in him – everything about him was utterly straightforward.

The credit, therefore, must go to his colleague and co-founder, the serpentine Reverend Manasseh Cutler, pastor of the Congregational church in Ipswich, Massachusetts – 'stately and elegant in form, courtly in manners', as an acquaintance described him, 'and at the same time easy, affable, and communicative. He was given to relating anecdotes and making himself agreeable.' It was Cutler, exquisite in black velvet suit, white linen collar and silver-buckled shoes, who convinced Congress to disregard the democratic principles it had enshrined in the 1785 Land Office Ordinance, with its townships and sections. It was Cutler who persuaded them to sell the Ohio Company over a million acres immediately, with an option on almost four million more. His task was made easier by the slowness of Hutchins' survey of the Seven Ranges and Congress's impatience to make money quickly from land sales, but as Jeffrey L. Pasley argues in his essay 'Private Access and Public Power' (1998), the manipulative, charming Cutler embodied all the qualities of the army of fixers, dealers and lobbyists who were to follow him over the centuries through the corridors and offices of the United States government.

The government's price for the land was nominally $1 an acre, but Cutler's persuasive manner, helped by the occasional backhander, brought it down by a third, then by a half, and eventually to twenty-five cents an acre, most of which was paid in devalued bounty warrants, so that the actual cost to the Company was approximately twelve cents an acre. On the day that the deal was struck, in July 1787, the entry in the reverend's journal was a decidedly unspiritual salute to 'the greatest private contract ever made in America'. If Cutler takes credit for the commercial lobbying, Putnam was responsible for the more idealistic parts of the legislative framework within which the Ohio Company was to work.

On 13 July 1787 'An Ordinance for the government of the Territory of the United States northwest of the River Ohio' was passed by Congress. Between three and five states were to be carved from the territory, which extended as far west as the Mississippi. Their citizens would be guaranteed elected government as in the existing thirteen states, but, three years in advance of the United States Constitution, they also were assured of freedom of religion, trial by jury, and due process of law. At the last moment a clause was added requiring section 16 in each township to be set aside for the support of public education, on the grounds that 'Religion, morality and knowledge being necessary to good government and the happiness of mankind, schools and the means of education shall forever be encouraged.' As a result of Putnam's own firm convictions, slavery was also banned, almost eighty years ahead of its prohibition by the thirteenth amendment to the Constitution.

Much of the rest of this remarkable document bore the imprint of Jefferson's 1784 report and his vision of the way American democracy should develop. Consequently it began with a section which effectively made feudal land tenure and laws of primogeniture and entail illegal in the territory northwest of the Ohio. So that nothing should impede the flow of land from the central government to its new owners, it stated specifically that the territorial legislature 'shall never interfere with the primary disposal of the soil by the United States in Congress assembled, nor with any regulations Congress may find necessary for securing the title in such soil to the *bona fide* purchasers'. Finally all those exercising authority in the territory, its governor, judges, elected representatives and voters, were required to be landowners themselves.

Clearly much of Thomas Jefferson's outlook was shared by the New England founders of the Ohio Company, but what was being devised in the Northwestern Territory went beyond the New England experience. It needs to be emphasised what an extraordinary invention had been set in place. The hypothesis of Jefferson's experiment was that land ownership shaped society. In Europe it created a pyramid that descended from a ruler or ruling group through grades of power to the landless and powerless. Although the American colonists had objected to the demands of feudal land tenure, they found it difficult to escape the sense of social hierarchy that it imbued. Consequently, in their newly independent nation it seemed reasonable for John Adams to suggest that the first president should be addressed as 'Your Majesty', and for George Washington to adopt the trappings of monarchy, and for Jefferson to suspect everyone around the President of being 'monocrats'. What other form could a civilised society take?

Now, in the Northwestern Territory, perhaps for the first time in history, a society was to be created where land ownership was to be horizontal rather than vertical. If Jefferson's hypothesis was correct, the social structure that grew from it must be democratic.

Fixer and dealer that he was, Manasseh Cutler recognised this as the crucial attraction that the Territory offered, and, in the Ohio Company's prospectus, presented it in characteristically seductive fashion: 'There will be one advantage which no other part of the earth can boast and which probably will never again occur – that in order to begin *right*, there will be no *wrong* habits to combat, and no inveterate systems to overturn – there is no rubbish to remove before you can lay the foundation.'

In the spring of 1788, Rufus Putnam led the first party of forty-seven settlers to this new land. Having built a great barge, which they named the *Adventure Galley*, and several smaller boats, they launched them on the Youghiogheny river in western Pennsylvania, and drifted on the current until they met the Ohio at Pittsburgh. From there, they floated silently downstream, past the clearing in the forest made by Thomas Hutchins' surveyors, until on a dank, foggy morning a lookout suddenly saw on the right bank the dim outline of Fort Harmar. This marked the mouth of the Muskingum river, where the Company's purchase began. Frantically they rowed and poled for shore, but the river carried them past, and the party's first landing was some distance below the intended spot, forcing them to haul the *Adventure Galley* back upstream from the bank. The plaque which today commemorates the place where the settlers first set foot on shore quite properly overlooks this mishap, and places the start of the great experiment at Marietta itself, where the Muskingum joins the Ohio.

That it was an auspicious spot was suggested by the presence of tall mounds, part of the Hopewell civilisation, constructed there centuries earlier by Native Americans. Between the mounds and the Muskingum, and about three-quarters of a mile up from the Ohio, Rufus ordered the construction of a wooden fortress, surrounded by a palisade, which he named after the god of war, Campus Martius. The settlers were right to be wary. A party of Delaware who camped nearby in the long clover and buffalo grass while they traded buffalo and beaver pelts with the soldiers in Fort Harmar served as a reminder that not everyone believed this to be Ohio Company land.

The balance of power within the Company can be deduced from the changing name of the town that was planned. It began as Castrapolis, 'the armed city', which was Rufus's choice, and for obvious reasons was quickly replaced by Cutler's more agreeable selection, Adelphia, or 'brotherhood'. This was discarded when the Company's secretary and former surveyor, Winthrop Sargent, suggested that French immigrants might become purchasers, and this commercial consideration led to its final name, Marietta, a tribute to the French Queen, Marie Antoinette. The city was quickly laid out on the banks of the Muskingum.

The original design, drawn in the calm of Massachusetts, was for sixty large squares with sides 120 yards long, each square containing twelve house lots; but these were awkward dimensions for Gunter's chain to measure out on the ground. In the town plan of 1837, the squares have become oblongs '12 chains and 28 links in length and 5 chains and 60 links in breadth including the alley which is 16 links wide', suggesting that speed required the original plan to be adapted. Rufus's plan imaginatively incorporated some of the Hopewell mounds into the town's squares, but in those first years the settlement was plain rather than beautiful, and the rows of log cabins had the look of a military barracks.

In 1787 Rufus sent out four survey parties armed with muskets, compasses and Gunter's chains to lay out the Company's claim into townships and lots in accordance with the Land Ordinance. Since the claim was situated next to the Seven Ranges, its squares were supposed to align with those run by the public survey, and like Hutchins' men the survey teams caterpillared their way chain by chain through the forest, producing results no more accurate than theirs. Although the north–south township lines roughly matched with those of the public survey, the range lines running east–west missed so badly that the map looked like a staircase. Added to the fears of Indian attack, the shoddy survey work slowed sales so much that Rufus had to take time out to persuade Congress to defer payment of the next instalment of the price. It was not a job for which he was suited, and for probably the first time in his life his courage failed. A frantic letter arrived in Ipswich, Massachusetts addressed to the Reverend Cutler.

'Your presence here is very much wanted,' Rufus wrote. 'I have not time to explain to you the necessity of your being on the spot as soon as possible, can only tell you that if you have any regard to your own interest, you will set out for New York without loss of time – I repeat it my dear friend come on without loss of time – Come my dear friend don't fail.' The silver-tongued cleric raced to Rufus's assistance, and won the Company a temporary reprieve.

Before the end of the year the first law court was established at Marietta, the first governor, Arthur St Clair, was installed, and the first families had arrived. In November 1790 Persis came with her furniture and surrounded by the Putnam family, now numbering eight children and two grandchildren. And that year President George Washington paid the new settlement a glowing compliment. 'No colony in America was ever settled under such favourable auspices as that which was first commenced at the Muskingum,' he announced. 'I know many of the settlers personally, and there never were men better calculated to promote the welfare of such a community.'

The one Washington knew best was the settlers' leader, Rufus Putnam, and in his autobiography Rufus admitted modestly that whatever troubles the world might heap upon his straight back, his commander-in-chief's opinion of him was 'no small sorce of consolation'.

SEVEN

The French Dimension

IN THE AUTUMN OF 1784, Rufus's Arch Enemy arrived in
Paris as the Minister of the new United States to France.
Thomas Jefferson's entourage included his daughters, slaves,
cook, secretary and valet, so his first duty was to find a good-sized
residence to serve as both home and embassy. He eventually
settled on the Hôtel Langeac, a building in a new development
between the Champs-Elysées and the rue Saint-Honoré.

Five years earlier the site had been a nursery garden belonging to the royal family, but the King's brother, the comte
d'Artois, had decided it was more valuable as real estate, and
had it measured, divided into lots and sold off for housing.
In doing this, he made himself simply the most prominent of
more than a dozen dukes, marquises and counts who had
joined forces with architects and developers to create a real-
estate boom that was transforming the French capital. Every
day new buildings were going up on what had been the market
gardens that fed Paris, pushing the city limits out towards
Montmartre and the old parade ground of the Champ de Mars.

What was happening in Paris could also be found in other
more developed parts of France like Lyon and Bordeaux. In
the course of the eighteenth century, over 150 years after
England, and almost a century after the American colonies,
France had started to discover the rapturous delight of property. Common land was enclosed, woodland fenced off,

grazing rights taken into private hands, and aristocratic estates sold to wealthy merchants. As always, the surveyors were the litmus paper that indicated the changeover from feudal to private ownership.

In Lorraine in eastern France, for example, an accelerating programme of enclosures was putting an end to the old open-field system of farming, and tracts of land were being engrossed into single fields. Instead of the old peasants' measures based on the amount of ground needed to feed a family, the *geomètres* or surveyors used a precisely defined *perche*, consisting of eighteen *pieds*, each of which measured exactly 12.789 inches. One hundred square *perches* made an *arpent*, and the area could all be drawn to scale on a map, and defined as the absolute property of a single owner.

In Saint-Etienne, near Lyon, the old variable units were also replaced, and the surveyors boasted that 'we have put all these defective [measures] in good order, so that in each district their content is regulated in either *perches*, *pas* [sixty-four inches] or *pieds*'. Striking a cautious note, they advised that it was best to consult the peasants before enclosing open fields, but the fact that the advice was needed suggests that the contrary often took place. The sort of defective measures they were replacing were those defined by the *cahier* produced in 1789 by the assembly of Bourges in central France: 'The *arpent* is not divisible by *perches* or *pieds*,' the assembly stoutly insisted, 'but by *journées* which means by fields that one man is able to plough in one day; according to local custom, one *arpent* is equal to sixteen *journées*.' The *journée* was the quintessential variable unit, equivalent to the daywork in England, the *Morgenland* in northern Germany, the *giornata* in Italy, or the *journal* in Catalonia. Wherever it continued to exist, so did the feudal economy in some form, and a true land market could not fully develop until it was superseded by exact measures of length and area.

In 1780, Alexi Paucton wrote in his now-classic work on

measurement, *Mètrologie*, about the superiority of the new in-variable measures over the old: 'They are the rule of justice and the guarantee of property which must be sacred.' It would be another fifty years before Pierre-Joseph Proudhon riposted, 'Property is theft,' but by then it was too late – the land had been stolen and property was everywhere.

As in Tudor England, in eighteenth-century France it was those in control of the measures who were able to establish their rights of ownership, while the others who had only feudal custom to justify their occupation of the land were dispossessed and forced to take to the roads looking for casual labour. In the early days of the Revolution, when the whole country was in a state of tension, the appearance of crowds of these wandering beggars frequently triggered rumours of invading armies.

It was not only in measuring the land that exact units were required. Factories and mines were becoming modernised and increasing in scale. New industries, like the de Gribeauval armaments concern which manufactured Napoleon's in-comparable artillery, had to have precise uniform measure-ments if components were to be fitted together and complex machinery was to function. Precision in one area encouraged it in another, for landed aristocrats invested in these new ventures, especially around Lyon where they owned some of the largest spinning mills, while industrialists bought them-selves titles and estates. A revolution was taking place long before 1789.

In her posthumously published memoirs, the conservative marquise de Crequy noted the social shockwaves of the late eighteenth century – the popularity of Mozart's *Marriage of Figaro*, which mocked the stupidity of the old aristocracy, and the way that 'people said "women" instead of "ladies", and "men of the court" instead of "Nobles". The greatest ladies were invited to supper and mixed with the wives of financiers.'

Even the most conservative members of the aristocracy tried

to take advantage of the improvements in agriculture which produced larger harvests and heavier animals. In cash-poor economies like western and southern France, landowners took their share of the growing profit by arbitrarily increasing the size of measures in which corn-rent and other dues were paid. That after all was the way feudal economics had always worked. The measures varied because in theory the amount of money was fixed – and in practice, cash was not readily available. To see it the other way round required a paradigm shift in thinking from ancient to modern.

Those in charge of the measures were not short of opportunities to increase their yield. 'The unending proliferation of measures is quite beyond imagination,' wrote Arthur Young, an agricultural expert who visited France in 1789. 'They differ not only in every province, but in every canton too, and almost in every town; the differences drive people to despair.'

The changes provoked an explosion of resentment that was broadcast to the world through the *cahiers de doléance* collected in 1788 from local assemblies across France. The opportunity to air their grievances arose when Louis XVI was forced to summon the Estates-General of nobles, clergy and the Third Estate – the democratic assembly which had last met in 1614 – to find some way of paying the nation's bills that had accumulated since the American Revolution. Members of the Estates were chosen by provincial and communal assemblies, who often also provided them with a mandate or catalogue of grievances which they wanted put right. It is in these *cahiers de doléances* that something can be felt of the banal, mundane, humiliating harm done to helpless people by arbitrary weights and measures.

On the one hand there were traditional proprietors increasing their measures, on the other cash-rich outsiders investing in a commodity on which they wanted a return and which gave them the means of doing so through control of local measures. The result was a howl of protest. Thus the parish of Limousis in Aude in the south complained that they had always paid rent with a measure called the *censuelle*, but that the demesne had been bought in 1774 by 'Sieur Rolland', who insisted on payment with 'the measure of Carcassonne', which was one-third larger. 'Everyone, anxious not to lose what little he possessed and reluctant to fall foul of his seigneur, yielded to the oppression and paid up in the way that he was coerced.'

From Brittany in the north-west the complaints were less specific but similar in tone. 'The measure at Plumaudan near Dinan has grown much larger,' declared the people of Yvignac near Rennes, and close by in Plancoët they protested, 'The measure they apply to the grain is arbitrary,' while from the rugged country near Quimper they insisted, 'The nobles' measure grows larger year by year.'

Years of suppressed indignation can be heard in a deposition from Lorraine, in the east of France: 'The lord of Corny extorts almost six hundred *hottes* of wine as rent. The amount used to be less, but his functionaries have been very clever at converting the *septiers* into *chaudrons* in such a way they gained a[n extra] quart for each *septier*. They cannot claim that this was unintentional because everyone knows that one *septier* of Lorraine has only four quarts whereas the *chaudron* has five.' This should have been a cause of embarrassment to liberals such as Thomas Jefferson, since Ethis de Corny and his wife Marguerite were part of his circle of progressive friends. But since the nobles' employees habitually skimmed their own share from the measures, such injustices were not rare. The accusation from Angoulême in the south-west was aimed at a similar trick, practised in this case by local millers who acted as the agents of the nearest lord: 'To collect the corn well-dried, to remeasure it after [it has become damp and swollen], to take the surplus that will thus accrue to every [bushel] for themselves and then to take a surcharge on the remainder as well – that is how they go about it.'

There are something like forty thousand of these dossiers, often ill-spelled, filled with clichés and windy rhetoric, and covering the entire range of rural woes. Many were evidently composed by the local schoolteacher or lawyer, and a few by aristocrats themselves, putting the best case for the seigneurs' position, but where weights and measures were concerned, the call for reform was deafening. 'We demand,' wrote a northern commune in the Pas-de-Calais, 'that the lords to whom we are obliged to pay corn tax be obliged to keep standards at the manor house, properly attested according to the old measure.'

References to 'the old measure' appear repeatedly, indicating how widespread was the anxiety at the changes being introduced. Before the present arbitrary, unfair array of weights and measures existed, the *cahiers* assumed there had been a single, uniform system – and that was what they wanted to get back to. As the Sens assembly put it, echoing the ancient policy of France's kings, 'Let there be in the Kingdom only one god, one King, and one law, one weight and only one measure.' Contemplating that lost simplicity, the frustration of the commune of Saint Sulpice-sur-Loire boiled over into a furious question that was implicit in every *cahier.* 'Why should the seigneurs and the privileged clergy enjoy the right to please themselves as to what measure of grain is binding on their estate?'

At least one man in France was confident that an answer could be found which would put the size of measures outside the waywardness of human control. In 1666 the Académie des Sciences had been established in Paris with the primary task of providing 'more accurate maps than we have hitherto possessed'. Since then one single project had dominated the Académie's activities – the mapping and exact measurement of France, a project that had forced its scientists to measure the world. The science and technology they had developed now made it possible to devise an entirely new system of measurement. For the first time in the ten thousand years since society began, its dimensions would not be derived from the human body or human activities, but from a mathematical abstraction. Certainty was needed, and science could provide it. A watershed had arrived.

That at least was the belief of Jefferson's most frequent guest at the Hôtel Langeac, the marquis de Condorcet, a mathematician and philosopher, passionate moderniser and idealistic advocate of Enlightenment values, who also happened to be the Permanent Secretary to the Académie. In the slang of the time, he was an '*idéologiste*', and since his wife shared his radical ideals, Paris gossip took particular pleasure when their only child was born in April 1790, enabling everyone to calculate that she must have been conceived on the most erotic night in radical history – 14 July 1789, when the Bastille fell.

Condorcet had written a book praising the democratic ideals written into the American Declaration of Independence, and with characteristic enthusiasm sought out Jefferson when the latter arrived in Paris, overcoming the natural barriers that might have been expected between two personalities so alike in their reserve and complexity. Not only did Condorcet share Jefferson's rationalism, they were the same age and resembled one another in their tallness, awkwardness and sandy hair, and in their divided temperaments. 'Between the shrewdness of his mind and the goodness of his heart,' observed Amélia Suard of her friend Condorcet, 'there was always a contrast that I found singularly striking.' The description could have served equally well for the American Minister.

The great project that Condorcet and the Académie were moving towards was that of giving measurement a scientific basis. This idea had been proposed as early as 1665 by the scientifically erudite abbé Gabriel Mouton at the Collegiate church of Saint Paul in Lyon. In a wonderfully creative leap, he suggested that either the size of the earth itself or the length of a second's pendulum – calculated by Newton to measure just over thirty-nine inches – might provide a solution to the problem of finding a unit of measurement that was invariable and universal and scientifically testable. Synthesising all the latest learning – Stevin's decimals, Galileo's pendulum, the concept of triangulation, the first estimate of the length of a degree of longitude – into one grand idea, abbé Mouton proposed that one minute of one degree round the circumference of the globe should become the universal base unit, to be called a *milliaire*. A new universal *toise* should equal one thousandth of a *milliaire*, its length would be checked against a second's pendulum, and its subdivisions, a universal *pied* and a universal *doigt*, would be a tenth and a hundredth respectively of the *toise*.

Mouton's idea was a freak. No one knows where it came from, and the knowledge to support the concept simply did not exist. Every practical ingredient in it was wrong – the shape of the earth, the dimension of a degree, and consequently the length of the universal *toise* – and as time went by and more was discovered about the complexities of measuring the exact size of an orange-shaped world, the less practical his proposal seemed.

Even to establish the length of the *toise*, France's basic unit of length, the problems seemed insuperable. In 1668 the Académie declared it be exactly six royal *pieds*, or 38.4 inches, and an iron bar of that length was prominently attached at eye-level to the wall of the great Paris fortress Le Grand Châtelet. The difficulties began to emerge the following year, when the foremost astronomers in Italy and France, Gian-Domenico Cassini and Jean Picard, started on the project for which the Académie had been founded, the mapping of France.

Like ordinary surveyors, they used triangulation, measuring a base-line and observing the angles, but not only did their measurements and observations have to be more precise, the size of the area they were surveying meant using spherical trigonometry to take account of the curvature of the earth. To aid them they had instruments of enormously increased accuracy. Looking through the telescope of an early-eighteenth-century theodolite, a mapmaker would be able to swivel it horizontally and vertically, line the cross-hair on a mark five or six miles away, note where the hair cut into a rotating compass divided into degrees, minutes and seconds, and expect an accuracy of about thirty seconds – over a mile this amounted to a possible error of about thirty inches, or around one in two thousand, small enough to be negligible. To keep errors from multiplying, the French surveyors checked their latitude and longitude from star-sights, and remeasured all the angles in the triangle to ensure that they added up to 180 degrees – a standby known as closing the triangle. Eventually a line of interlocking triangles and parallelograms which looked on paper like a girder bridge would stretch across the country. Triangulation was always a slow and painstaking process, but in France, the first country to measure itself accurately, it became a marathon of spectacular proportions.

Jean Picard measured his first triangle near Amiens in 1669, and went on working until his death fourteen years later, when the task was taken over by Gian-Domenico Cassini, who continued until he died, at which point the job fell to his son, Jacques Cassini, who in 1733 completed the meridian between Dunkirk in the north and Perpignan in the south, a line of triangles stretching from coast to coast. Fifteen years later, King Louis XV was presented with the *Carte géométrique de France*, covering 180 sheets, and caustically remarked that, having seen the true size of his kingdom, he had lost more ground to his mapmakers than he had won through conquest.

Royal doubts could be set aside, but science was another matter. For some reason, Jacques Cassini's triangulation had gone wrong. The length of France covers about nine degrees from north to south of the globe, and as he went further south, he found that the degrees of latitude appeared to become longer. What these measurements suggested was that the earth was elongated, shaped more like an egg than an orange – but according to Newton's theories of gravitation, it had to be orange-shaped. Such was the power of the Académie des Sciences that it persuaded the government to fund two hugely expensive expeditions to resolve the arguments. In 1735 one was sent south to Peru to measure the distance of a degree on the meridian – a line of longitude running north–south – at the equator, while a second went north to the Arctic circle to measure a degree of the meridian there.

What followed were years of furious dispute, and outright fraud as well as painstaking astronomical and terrestrial research in the frozen depths of Lapland and the snowy heights of the Andes; but in the end the results proved that the earth was shaped like an orange, that Jacques Cassini had made a mistake, and that France consequently needed to be re-surveyed.

Another team, under the cartographer abbé La Caille, began the task in 1738. His meticulous approach was of a new order, taking account of such tiny forces as the gravitational pull of the Alps, which swung his plumb-line minutely off the vertical. On La Caille's death, Cesar Cassini de Thury, the third member of the family, took over the triangulation, but in 1784 he died of smallpox, leaving a fourth Cassini, Jean Dominique, to supervise the final publication of the definitive map in 1789, 120 years after the project began.

No one could doubt that the French took their measurements very seriously indeed. Within the Académie, mapmaking and astronomy consumed the bulk of its annual budget of ninety-three thousand *livres* (about $18,000 or £4000), and scientists around the world acknowledged France's pre-eminent expertise in measuring the shape of the earth. What no one except a scientist could be confident about was the unit they used to estimate the distance of a degree. How long was a *toise*?

When the abbé La Caille began the second triangulation of

France using a *toise* calibrated against the standard at Le Châtelet, he discovered an unaccountable deviation in his figures not only from Cassini's, but from the utterly reliable Picard's. It was impossible to suppose that Picard would make a mistake, and the very consistency of the deviation, with Picard's *toise* always one-thousandth shorter than La Caille's, showed that Picard's must have been fractionally shorter than the iron bar at Le Châtelet. The Académie compared the different *toises* used in Peru, in Lapland and by La Caille, and in 1766 decided that the first of these would become the ultimate authority. Ignominiously, the Châtelet iron bar was hammered down until it measured the same length as the *toise du Pérou*, the new official standard.

At the same time, the Académie agreed on a new standard for the *livre* or pound weight, based on a pile of copper weights allegedly dating from Charlemagne's reign, and the government took the opportunity to send copies of these agreed standards of distance and weight to all eighty provinces in France. It was a gesture towards national consistency, but since the measures used in most markets continued to belong to the aristocracy, in practical terms it did little more than add a fresh possibility to the dozen existing lengths for the *toise* and the twenty-four or a hundred definitions of the *boisseau* and *quintal.*

By this time Mouton's suggestion had disappeared as a practical proposition. Had the earth been a perfect sphere, it would have been feasible to estimate its exact circumference, but working out how much an orange-shaped globe flattened at the poles was impossible, and the estimates varied too widely to be acceptable. Paradoxically, as instruments improved in the eighteenth century and measurement became more precise, it seemed ever less likely that an exact figure could be established for such a lumpy, uneven planet.

Scientists who were aware of the problem preferred the second's pendulum as the scientific basis for a new system of weights and measures. Picard and after him the mathematician La Condamine had both favoured the pendulum, and in 1775 when Condorcet was commissioned by the government to investigate the subject, it was no surprise that he should have come to the same conclusion. By then it was accepted that Newton's laws of motion were universally applicable, and the strongest mathematical and experimental proofs had come not from Britain, but from the work of brilliant French mathematicians like Joseph-Louis de Lagrange and Pierre-Simon de Laplace, and the heroic efforts of more than a century of French geodesy or earth-measuring.

As he had shown in 'Some Thoughts on a Coinage', Thomas Jefferson was already familiar with much of this intellectual background, but after his meetings with Condorcet in Paris he too became convinced that it was impossible to measure the earth accurately, and his 1784 proposal for a new decimal system based on the length of the equator disappeared from sight. Instead he opted for the second's pendulum, and he quickly convinced his friend James Madison of its merits. Madison in his turn wrote to James Monroe, his successor as President, in 1785, urging that it would be 'highly expedient, as well as honourable to the federal administration, to pursue the hint which had been suggested by ingenious and philosophical men, to wit: that the standard of measure should be first fixed by the length of a pendulum vibrating seconds at the Equator or any given latitude'.

It was obvious to both Jefferson and Condorcet that science and democracy were natural allies. 'A good law ought to be good for all men,' went one of Condorcet's maxims, 'as a true [scientific] proposition is true for all men.' Finding a scientific basis for measurement would remove control of weights and measures from the privileged and powerful, and make it available to anyone who knew arithmetic or could consult a scientist. In words that Jefferson would echo in the United States, Condorcet asserted that 'The uniformity of weights and measures cannot displease anyone but those lawyers who fear a diminution in the number of trials, and those merchants who fear anything that renders the operations of commerce easy and simple.'

In August 1789, a month after the Third Estate succeeded in establishing itself as the National Assembly of France, a committee was appointed to consider the question of aristocratic and Church privileges. These included feudal taxes like Church tithes, the *gabelle* or salt tax and the *corvée* or monthly labour on the roads, as well as the right to use their own weights and measures. History takes the French Revolution's outstanding date as 14 July 1789, when the symbol of despotism, the Bastille, fell – but for the great mass of the population the crucial moment came on 19 March 1790, when the Assembly voted for the abolition of aristocratic privileges, and at a stroke the power that underpinned the whole structure of local government and social life in the *ancien régime* was swept away.

With the privileges disappeared the old standards – literally. In May 1790, when the National Assembly ordered parishes to report what weights and measures were used locally, most of them were unable to comply, because the aristocrats had taken the actual scales and containers away. 'We have no standard to verify our weights and measures,' one district confessed, 'for the practice was for the seigneur alone to see to this.' 'The standards for our measures are the property of the former duchesse de Corge,' explained the parish of Mathay in the modern department of Charente. 'We ourselves are unable to present any documents'.

A vacuum had been created, and the scientific community was aching to fill it. In March 1790 Condorcet persuaded the Bishop of Autun, Charles Maurice de Talleyrand-Périgord, best known simply as Talleyrand, to submit a proposal to the National Assembly calling for a new system of weights and measures based upon the second's pendulum.

The unifying theme in all the National Assembly's deliberations was Rousseau's axiom that men were born equal and were divided only by the hierarchies thrown up by a fundamentally irrational, old-fashioned society. Once these arbitrary barriers were demolished and replaced with new institutions based on reason, the essential fraternity and equality of mankind could flourish. In this spirit, the committee on government proposed sweeping away the crazy overlapping pile of tax, military, judicial and aristocratic jurisdictions and basing the government of the country on a grid which bore a striking resemblance to Jefferson's plan for the United States's western lands. Where Jefferson wanted square counties subdivided into square hundreds and square lots, the National Assembly's committee would have partitioned France into eighty square departments, each measuring eighteen by eighteen leagues, divided into nine square districts, and each district subdivided into nine square communes. In the Assembly, however, local prejudices and traditions made themselves felt, and France ended with an arrangement of eighty-three departments of almost every shape other than square.

There was good reason to suppose the same fate would befall any radical scheme presented to the Assembly for the reform of weights and measures. All that the *cahiers de doléance* had asked for was that the same set of measures should be used across the country, not for a wholly new system.

Condorcet's plan for uniformity based on the second's pendulum went first to the committee on agriculture and commerce, which reported that uniformity was indeed desirable for both farmers and shopkeepers. In May 1790, the Assembly adopted the committee's report and gave the Académie the responsibility for deciding on the best way of achieving uniformity. At the same time it asked the King to invite his brother-monarch, George III of England, to co-operate in the venture, and particularly to arrange a meeting between Britain's foremost scientific institution, the Royal Society, and

the Académie 'to determine the length of the second's pendulum, and to derive from it an invariable standard for all
measures and weights'.

For the scientists appointed by the Académie to be on the
commission for reform of the measures, Jean-Charles Borda,
Joseph-Louis Lagrange, Pierre-Simon Laplace, Mathieu Tillet
and Marie-Jean Condorcet, this was a moment of triumph
when science and democracy came together. 'Everything tells
us that we are approaching the period of one of the greatest
revolutions of the human race,' Condorcet wrote in his most
utopian mood. 'The present state of enlightenment guaran-
tees that it will be happy.'

Condorcet was extreme, but all the French scientists seemed
to sense the intellectual grandeur of the enterprise. To devise
a new system of weights and measures on a scientific basis
involved a leap of imagination from the human-centred to the
abstract. It had no precedent in history. It was like inventing
a new colour. It was, said Lavoisier, disregarding his own pion-
eering discovery of the properties of oxygen and the compo-
sition of water, 'one of the most beautiful and vast conceptions
of the human mind'.

Any politician who believed in democracy but secretly saw
himself as a scientist was bound to be intoxicated by such a
project. Nevertheless, it was an extraordinary coincidence that
in the summer of 1790, after ten thousand years of communal
living, the beautiful and vast conception of reinventing the
world's weights and measures should have been duplicated
across the Atlantic. Not surprisingly, there too the new system
was based on the second's pendulum.

EIGHT

Democratic Decimals

IT WAS ONLY THE NEED to settle his teenage daughter in an American school that took Thomas Jefferson away from the congenial company of French scientists in September 1789. He returned to find a United States with a new Constitution and Bill of Rights, and a new President who insisted that Jefferson take up the post of Secretary of State in the new federal government. When George Washington for the first time addressed the Congress created by the Constitution on 2 January 1790, he outlined with some care the most pressing matters that he felt they should consider for action. The first two were defence and the economy, and immediately after them came the need for a uniform system of weights and measures. Acting on this suggestion, the House of Representatives requested the new Secretary of State on 15 January to draw up 'a proper plan or plans for establishing uniformity in the Currency, Weights and Measures of the United States'. Consequently the first job waiting for Thomas Jefferson when he arrived in New York on 15 April to take up office was one close to his heart.

In May, soon after Jefferson had begun work, Congress passed a bill to create 'The Territory of the United States South of the River Ohio', which consisted of lands once claimed by Virginia and South Carolina but which were now the Territories of Kentucky and Tennessee. Governors were appointed, boundaries defined with the promise of a detailed survey to come, and the timetable of evolution to statehood, as proposed by Jefferson in 1784, was laid out. Already settlers were pouring into the region, mostly Scotch-Irish and Germans coming down the Shenandoah valley from western Pennsylvania and the Cumberland Gap from Virginia. Together with the spread of migrants into the Northwestern Territory and the newly created state of Vermont, this expansion of the United States underlined the urgency of Jefferson's work.

The settlers brought with them their gallimaufry of measures, their Rhineland *Ruthin* and Scots miles and Irish acres, their mutchkins, flitches, fardels, *liepsfunds*, *quentchens*, hattocks, thraves and *zehnlings*. For genuine frontier families who lived off what they raised, the only goods that needed measuring by strangers were the ones they had to buy – flour, coffee and sugar in market, and cloth from peddlers. Among themselves they used their own lengths and capacities. In western Pennsylvania these tended to be Rhineland measures, in Vincennes on the Wabash river they were French, in New York, Dutch; but no matter what their nationality, country dwellers knew that when they came to town, the scales were literally loaded in favour of the merchants who owned them.

The problem had begun with the arrival of the original colonists. Virginia made its initial attempt to legislate against fraudulent measures within a generation of the first plantation, a second in 1646 when legislators accused both Dutch and English merchants of practising 'deceit and diversity of weights and measures', came back for a third try in 1661 with an Act whose preamble began, 'Whereas dayly experience sheweth that much fraud and deceit is practised in this colony by false weights and measures . . .', and followed this up with more legislation in 1734 – their fourth attempt in a century, accompanied by the sad explanation, 'Forasmuch as the buying and selling by false weights and measures is of late much practised in this colony to the great injury of the people . . .'.

In fact the phrase 'frauds and deceits' had appeared so often by then that it had become a cliché. It was used in the preamble to a Maryland law of 1671 referring to 'much fraud and deceit [that has been] practised in the province by use of false weights and measures'. An eighteenth-century North Carolina law used it in the plural – 'Whereas many notorious frauds and deceits are daily committed by false weights and measures . . .' – as did the New Jersey Act of 1725 calling for 'one just weight and balance, one true and perfect standard for measure, for want thereof experience has shown that many frauds and deceits have happened'. In New York, where the range of English measures was complicated by the continuing existence of Dutch units, and by the tendency of unscrupulous retailers to buy cloth by the English ell, which was almost a yard, but sell by the Flemish ell, barely half its length, the law plaintively pointed to the need for 'one true and perfect standard and assize of measure among them; for want whereof experience shews that many frauds and deceits happen, which usually fall heavy upon the meanest and most indigent sort of people, who are least able to bear the same'.

For all its democratic assemblies, its fine Constitution and splendid Bill of Rights, down in the marketplace the new United States often looked remarkably like the old colonies. The legal bushel in Pennsylvania, for example, was supposed to be wide and shallow, so that the amount of grain heaped above the rim would equal one-third of the amount in the container itself. In Philadelphia, however, shopkeepers made the bushels deep and narrow, so that when filled they would only take a small heap above the rim.

In Maryland, farmers tried to take court action against merchants who combined to introduce a larger measure for use in buying grain, but the justices of the county courts, who tended to be urban citizens in good standing, like shopkeepers, millers and grain merchants, were also responsible for weights and measures. And so, as a contemporary critic,

John Beale Bordley, noted, the merchants simply appointed sympathetic justices whose job was to 'order new standards, which had great weight in quieting opposition to the new half bushel measure'.

It would be hard to imagine a problem more attractive to Thomas Jefferson's superlative combination of lucid intelligence and fierce emotion. The typical victims of false measures were those small farmers on whom the health of democratic government depended, but who were easy prey when they came to market to sell their produce and stock up on necessities like cloth and sugar. The abuse of those Jefferson most admired by the urban traders he most hated was an affront to what he termed 'the democratic principle', and was thus intolerable. Because control of weights and measures was a source of profit, it would always end in the hands of the powerful. The only sure recourse was to make the system so transparent that 'the whole mass of the people . . . would thereby be enabled to compute for themselves whatever they should have occasion to buy, to sell, or to measure, which the present complicated and difficult ratios place beyond their computation'.

The issue was clear, but clarifying it further was the presence among the opposition of Jefferson's old nemesis, Robert Morris, who refused to believe that change was necessary. Their conflict had moved on in the six years since the defeat of Morris's bid for a new currency and, closer to his speculative heart, the chance of running the United States Mint. In 1784, while Jefferson was in Paris, Morris had come out ahead by negotiating, behind the American Minister's back, an exclusive agreement with the French tax-collecting authority, the Farmers-General, to sell American tobacco to France. Two years later, Jefferson hit back by persuading the French government to buy tobacco from other American, and especially Virginian, suppliers. Now both men were in New York, Morris riding a rising tide of land speculation and a soaring market in government securities, and Jefferson determined to give republican farmers an even chance against the chicanery of city sharks, shopkeepers, milling cartels and land speculators.

The task he had been given by Congress ought, therefore, to have been congenial, especially as it took him back to the mathematical problems with which William Small had wooed him into Enlightenment thinking. And he admitted as much in a letter to the astronomer David Rittenhouse: 'Five and twenty years ago I should have undertaken such a task with pleasure because the sciences on which it rests were then familiar to my mind and the delight of it. But taken from them . . . I have grown rusty in my former duties.'

Jefferson was handicapped by a lack of reference books – he had expected to return to Paris, and most of his library was still there, with the remainder in Monticello. But the chief drawback was that the Secretary of State hated being in New York. It represented everything he most disliked about cities, being full of noise, odours, and people on the make. During the Revolution, the British had bombarded it, the Americans had thrown up barricades, fires had broken out, and buildings had collapsed. By the time peace came, weeds grew between Broadway's cobbles; but nothing could alter the magnificence of New York's harbour, and at once an orgy of rebuilding began. Traders returned, and they were accompanied by the financiers and shipbuilders who made their business possible. The presence of Congress meeting in the City Hall on Wall Street brought in printers and publishers, and gave trade to coffee-shops, boarding-houses and restaurants.

When Jefferson arrived in April 1790, the reborn New York was overflowing its pre-Revolution boundaries at the foot of Manhattan. Jefferson's house in Maiden Lane in Lower Manhattan, several blocks north of Wall Street, should have been on the outskirts of the city; instead it had been engulfed by the thunder of construction as new red-brick houses were built for the expanding population. New York, said Jefferson with the moderation to be expected of a rational man, 'is a cloacina [shit-hole] of all the depravities of human nature'. Not only was he shut up in the cloacina as spring merged into the oven-heat of summer, he had set himself a Herculean task – to invent from new an entire system of measures.

He had, however, one invaluable colleague, in the shape of James Madison, for whom reform had become no less of a crusade. Among Madison's many quiet qualities was a tireless capacity for research, and it was he who found a pamphlet by Robert Leslie, a Philadelphia watchmaker, who recommended a solid rod for measuring seconds instead of a pendulum. Jefferson adopted the idea without hesitation. Using a rod removed the difficulty of measuring the exact middle of the bob at the end of a pendulum, what he called 'the point of oscillation'.

Jefferson's conversations with French scientists were evidently fresh in his mind, and he must have had available his 'Thoughts on a Coinage', because with astonishing speed he was able to sketch out the principles of the new system and its major divisions by the middle of May. On one page which started out as a fair copy of his initial findings, he wrote, 'Sir Isaac Newton has determined the length of a pendulum vibrating Seconds in latitude [unable to remember the figure of 51 degrees 31 minutes – the latitude of London – he left a blank space until it could be checked] to be 39.2 inches = 3.2666 &c feet measuring from it's point of suspension to it's center of oscillation. A rod vibrating seconds must be of the same length between the point of suspension and center of oscillation: and this center will always be found at two thirds of the whole length. Such a rod then will be 58.8 inches or 4.9 feet English measures long.'

A unit of almost five feet did not fit with any of the old measures, but doubling the length brought Jefferson close to ten feet. One-tenth of this length gave a decimal foot which would be near enough to the old foot to make it easy to introduce. 'Let the standard of measure be the Double length of a rod or Treble length of a Pendulum vibrating seconds,' he wrote decisively. 'It will yeild [*sic*] us the following series of divisions and multiples.'

What started as a fair copy was then gradually defiled by crossing-outs and tiny insertions of new figures and different assumptions as Jefferson attempted to juggle three incompatible assumptions – that the new units should be close to the old, that they should be decimal, and that weights, lengths and currency should all be integrated.

At the beginning of May, after three weeks' work, the stress of this intense and fiddly computation carried out in the noxious surroundings of New York became too much for him, and he was struck by an intense migraine. Years later he remembered it as 'a severe attack of periodical head ach [*sic*] which came on every day at Sunrise, and never left me till sunset'. At its worst, it forced him to sit in a room too dark for writing and to work out problems by mental arithmetic. 'What had been ruminated in the day under a paroxysm of the most excruciating pain was committed to paper by candle-light, and the calculations were made.'

Gradually the severity of the attacks wore off, and Jefferson returned to daytime work. Less than two months after he received notice of Congress's request, he had completed the task. It was, and remains, a formidable feat of intellectual application.

He began by choosing the swing of Leslie's rod for the base unit: 'To obtain uniformity in measures, weights and coins, it is necessary to find some measure of invariable length with which, as a standard, they might be compared.' The size of the earth might have served, but Condorcet had convinced Jefferson it was impossible to measure the equator or the whole of a meridian accurately. On the other hand, Jefferson pointed out, 'The motion of the earth round it's axis is uniform and invariable.' This produced a day and a night of varying duration, but averaged over a year the length of a day was everywhere the same, and could be divided into 86,400 parts, each of which was known as a second. The length of a pendulum, or better still a rod, which took one of those seconds to swing from one end of its arc to the other, could also be established. 'Let the Standard of measure then be an uniform cylindrical rod of iron, of such length as in lat. 45° in the level of the ocean, and in a cellar or other place, the temperature of which does not vary thro' the year, shall perform it's vibrations, in small and equal arcs, in one second of mean time.' In his first draft, Jefferson chose thirty-eight degrees north, the median line of latitude running through the United States, as the point where the pendulum or rod should be measured, but he changed it to forty-five degrees, the latitude of Paris, to harmonise with the French proposals.

Based on the second's rod, Jefferson then offered two solutions, under the pretext that he was not clear how radical a plan Congress wanted. Did they want a root-and-branch reform, similar to his decimal dollar system in place of the old pounds, shillings and pence? 'The facility which this would introduce into the vulgar arithmetic,' he explained, 'would unquestionably be soon and sensibly felt by the whole mass of the people who would thereby be enabled to compute for themselves whatever they should have occasion to buy, to sell, or to measure, which the present complicated and difficult ratios place beyond their computation for the most part.'

His own writings made it clear that this was what he himself wanted; but the best part of twenty years' experience in colonial, state and federal assemblies had taught him that the House of Representatives would need to argue themselves into supporting such a radical scheme, so he offered them an alternative: 'Or is it the opinion of the representatives that the difficulty of changing the established habits of a whole nation opposes an insuperable bar to this improvement?'

Because the first settlers came mostly from England, the weights and measures generally used in the United States were English, and, Jefferson explained with barely concealed disdain, 'We must resort to that country for information of what they are or ought to be.' In 1757, he noted, the House of Commons had appointed a committee under the Earl of Carysfort to examine the United Kingdom's weights and measures. It had found that the original seventeenth-century Exchequer standards constructed for Elizabeth I were 'brass rods, very coarsely made, their divisions not exact, and the rods bent'. It was true that in 1742 the Royal Society had had a new and very accurate yard standard made, but Americans could not know exactly how long it was because 'they furnish no means, to persons at a distance, of knowing what this standard is'.

This was not wholly honest – Elizabeth's yard despite its construction was astonishingly accurate, and it would have been easy to commission an exact copy of the Royal Society standard – but it served Jefferson's purpose to cut American measures free of the English standards. If units of distance like the foot and inch, and units of area like the acre, were based on the length of the second's rod, the existing units could be left unchanged, but they would be based on American science.

Then Jefferson turned to measures of capacity, and the sly turn of his humour can still be caught in the deadpan way in which he lists their absurdities:

> . . . 8 gallons make a measure called a firkin in liquid substances, and a bushel, dry;
> 2 firkins or bushels make a measure called a rundlet or kilderkin, liquid, and a strike, dry;
> 2 kilderkins or strikes make a measure called a barrel, liquid, and a coom, dry, this last term being antient and little used;
> 2 barrels or cooms make a measure called a hogshead, liquid, or a quarter, dry . . .

Having listed every last one of them, he delivers the punch-line: 'But no one of these measures is of a determinate capacity.' The contradictions in English law meant that each of them could be defined in at least eight and as many as fourteen different ways.

Having demonstrated the measures' unreliability, he discarded them all, beginning with the wine gallon, because it was only important to 'the mercantile and the wealthy, the least numerous part of society'. He proposed a new gallon of 270 cubic inches, which was roughly halfway between the most commonly used measures. All other capacities would be derived from it.

Then, with that scalpel clarity that enabled him to cut through constitutional and political complexity, Jefferson disposed of a problem that had caused complaints for centuries: 'The measures to be made for use [shall be] four-sided, with rectangular sides and bottom . . . Cylindrical measures have the advantage of superior strength, but square ones have the greater advantage of enabling every one who has a rule in his pocket to verify their contents by measuring them.'

It was the simplicity of the square that made it democratic. As Jefferson explained, anyone could measure it. The same reasoning had led him to choose it for the survey of United States public land, because, in the words of the geographer William D. Pattison, 'rectilinear land boundaries put it in the power of any settler, employing the most rudimentary means of measurement, to verify the contents of his purchase'. But it is possible to guess that the matter went a little deeper in the Jeffersonian psyche.

Consciously or otherwise, he kept proposing the square as the solution to all sorts of problems. Administratively, he hoped that Virginia's counties would be square, so that they could be subdivided into square wards. 'The wit of man,' he declared, 'cannot devise a more solid basis for a free, durable and well administered republic.' Planning the nation's new capital city in 1792, he imagined it as a ten-mile-broad square, divided into smaller squares, whose functions he outlined in a letter to George Washington: 'For the Capitol and offices, one square. For the market, one square. For the public walks, nine squares consolidated.'

The square offered not just simplicity but symmetry, and when Jefferson came to make his imaginative constructions real, that is what chiefly emerges. Thus the ground-plan of Monticello, the house he designed and spent forty years 'pulling down and putting up', and where his heart always remained, reveals itself to be a central oblong, consisting of the hall and rotunda, which separates two balancing squares. The campus of the University of Virginia, the greatest pride of his later life, is a large square enclosing two oblong rows of pavilions and gardens symmetrically arranged on either side of a lawn. Even in unconscious details – like the *maison carrée*, the rectangular Roman building in Nîmes which Jefferson selected as the original for Virginia's capitol in Richmond, or the four-sided *partie quarrée* he remembered with such affection from his youth where he and William Small, George Wythe and Francis Fauquier had set the world to rights – the square's perfectly balanced shape kept reappearing.

And the balance in the square was also intrinsic to Jefferson's vision of democracy. As he made plain in a letter to James Madison from Paris in 1787, what he liked about the United States's proposed Constitution was the separation it made between the three elements of executive, legislature and judiciary; but he insisted that it needed a fourth side to balance it, a Bill of Rights for the people. The shape ensured that American politics would always consist of a debate between the four parties – giving later generations the chance to hear a raucous echo of those Williamsburg discussions on which the young Jefferson cut his intellectual teeth.

Thus the squareness of the containers was not just a detail in Jefferson's scheme of things. It was an integral part of the pattern that he wished to impose upon the whole structure of his ideal state. To rule out the smallest room for doubt, he specified the dimensions of each container and required its contents to be measured striked, that is level with the rim, rather than heaped.

Jefferson's proposals for weights were also aimed at simplification. He wanted to amalgamate the two existing systems – the commonly used avoirdupois, whose range extended from drams to tons, and the specialist Troy weights used mostly by jewellers and apothecaries. The basic unit of this unified system, which was essentially avoirdupois, would be the ounce, weighing exactly one-thousandth of a cubic foot of rainwater.

It was a wonderfully simple proposal, but one designed almost entirely to set the House of Representatives thinking about the need for reform. Once it had succeeded in that object, Jefferson was ready to spring the more serious option on them. 'But if it be thought,' he continued, 'that, either now, or at any future time, the citizens of the U.S. may be induced to undertake a thorough reformation of their whole system of measures, weights and coins . . . greater changes will be necessary.'

In less than a thousand words, he then outlined the first scientifically based, fully integrated, decimal system of weights and measures in the world. Its basic measure of length, derived from the second's rod, was a foot, which would be divided into ten inches. A cube of rainwater, whose sides were one decimal inch long, was to weigh one decimal ounce, and ten of these ounces would make a pound. The basic unit of capacity would be the bushel, which was to measure one cubic foot, that is to say one thousand cubic inches. Finally, the weight of the dollar was to be adjusted so that it came to exactly one decimal ounce.

The system was plain and elegant, and the manner in which the three dimensions of weight, length and capacity were integrated, anticipating one of the great strengths of the metric system, gave it enormous coherence. None of its elements was original to Jefferson, save only his insistence that the bushel and other measures be 'four-sided and the sides and bottom rectangular', but the synthesis of Newtonian physics with the democratic aim of 'bringing the principal affairs of life within the arithmetic of every man who can multiply and divide plain numbers' was entirely his creation.

It was a remarkable moment in intellectual evolution. Measurement consists of abstracting one quality – length, mass, time, velocity – from an object or an event, and giving it a numerical value. Originally the unit used to make the measurement might have been personal – 'your' foot; then it became social – 'the' foot, an agreed average; now the foot had evolved into something scientific – a fraction of time.

The timing of Jefferson's report was critical. Any proposal for reform would encounter not only the outright hostility of traders, but the passive resistance to change of any kind always felt by the majority of the population. There was one exception to this general rule. Out on the frontier, Germans, Scots and Irish settlers, all reared with their own ideas of the distance of a mile, or the area of an acre, or the length of a rod, were quickly adapting to the basic reality of Gunter's twenty-two-yard chain, to the 4840-square-yard acre it measured out, and to the square-mile lot that made up the thirty-six-square-mile township.

English settlers who had moved to New Orleans or west of the Mississippi were equally adaptable. When it came to acquiring land, they were prepared to measure it as happily in French *arpents* and Spanish *labores* as in acres, and could learn to estimate a river-frontage in *toises* and in *varas* without difficulty. By the best estimate, in 1795 over one million acres, or almost six thousand *labores*, in Spanish America were owned by families brought up on English land measures, and a stream of others were moving into the rice-growing, timber-producing, French-measured land around the Gulf of Mexico.

Jefferson's first attempt to link his new measures to the sale of public lands in 1784 had failed, but his instinct was sound. The future of decimalised measures was tied to that part of the United States where measurement was critical, and where there was a clear willingness to accept strange and novel units. That was a view accepted by the Secretary of the Treasury, Alexander Hamilton, the only other member of George Washington's administration to rival Jefferson's stature.

With the hindsight of history, it has been customary to see Hamilton and Jefferson as natural opponents – pragmatic banker confronting political ideologist, urban striver opposed to landed aristocrat, action man against ivory-tower thinker. Certainly there was a bouncy, Tigger-like quality to Hamilton that might have been designed to rile Jefferson's Owlish pretensions, but at this stage they were colleagues and – almost – friends, who had seen at first hand the weaknesses of the Continental Congress, and were desperate to make the new Constitution's experiment with democracy work.

They had both trained as lawyers, and were imbued with the rational values of the Enlightenment. But while Jefferson was increasingly influenced by Rousseau's belief in the corrupting effects of wealth, Hamilton took on board the pragmatic Scots outlook of David Hume and Adam Smith that prosperity created civilisation. It was no coincidence that Hume found nothing noble in a society of small farmers – he had seen Scotland move from rural poverty to commercial surplus, and appreciated the difference. Like an Old Testament prophet, Jefferson warned that wealth poisoned society and that urban society was the most toxic of all; but Hamilton blithely accepted David Hume's argument that prosperity allowed arts and refinement to flourish, and nowhere more obviously than in cities. 'Thus,' ran Hume's theme, 'industry, knowledge and humanity are linked together by an indissoluble chain.'

Nevertheless, it was an indication of their relationship that on 20 June 1790 Jefferson should have brokered a meeting between Madison and Hamilton which brought a famous bargain – the location of the nation's permanent capital on the Potomac river, on Virginian territory, which Madison wanted, in exchange for the assumption of state debts by the central government, the project on which depended all Hamilton's plans for the restructuring of the United States's finances. Soon afterwards, Jefferson sent Hamilton a copy of his weights and measures report, and Hamilton replied in co-operative spirit. He had read the report with 'much satisfaction', he said, and endorsed fully the concept of 'a general standard among nations [which] seems full of convenience and order'.

Later in the autumn the first draft of Hamilton's own report on establishing a United States Mint appeared, and it contained a passage which would have warmed the Secretary of State's heart. Referring to Jefferson's proposal to make the weight of the dollar equal to a new decimal ounce, Hamilton wrote: 'There is an accuracy and spirit of system in this suggestion which constitutes a strong recommendation of it, if the advantage is not to be procured at the expense of some more considerable benefit. Fortunately there is such a coincidence of essential principles with this systematic idea, that it is not difficult to adhere to it.'

Perhaps the most remarkable instance of their harmony came in relation to the western lands, when Hamilton replied to a request from the House of Representatives for advice on ways to stimulate the so-far disappointing land sales. Unlike Jefferson, who saw the process as an experiment in social engineering, Hamilton considered it simply as a source of funds to help pay the nation's debts. But reporting to the House a month after receiving Jefferson's report on weights and measures, he extended an unexpected helping hand to the experiment.

Most of his suggestions were pragmatic: reducing the price per acre from $1 to 30 cents, allowing two years' credit, and establishing additional land offices so that purchases could be made not only in the capital, but where the land was, out on the frontier. But then came the gesture of support to Jefferson. The public lands should continue to be surveyed and laid out as a grid before they were sold, Hamilton reported, but in future the townships should be ten miles square, containing lots measuring one hundred acres and upwards. He went no further in specifying measures and dimensions, but evidently did not consider it necessary. Jefferson's report had already laid out the values of the decimalised units of length, capacity and weight that were being proposed for the United States. If Congress accepted those recommendations, the new units would necessarily be applied to the survey, and Hamilton's plan to decimalise the survey into tens and hundreds would neatly accommodate the change.

What each man implicitly recognised was that the measures used in the survey of the public lands would also become the standard for the country as a whole. For the moment the measures used were Gunter's, but any system might be planted in the wilderness. There was, in Reverend Cutler's lambent phrase, no rubbish to be thrown out.

Consequently Jefferson had grounds for confidence that his report would be well-received. The Constitution placed responsibility for weights and measures with Congress. They now had a coherent plan that had the powerful backing of the President, of Hamilton and Jefferson, the strongest voices in the executive, and the indefatigable support of James Madison, one of the legislature's most influential members. It was what the hostile Massachusetts *Western Star* called an 'overbearing and intriguing majority'.

In the country at large most people were probably indifferent, but there was a small but vocal section of opinion in favour. As the New York *Daily Advertiser* declared, 'This great desideratum in commerce and in social life . . . will probably be at length attained, and England, in conjunction with

France, will perhaps have the honor of conferring this benefit on the rest of Europe.' Pamphlets like John Beale Bordley's *On monies, coins, weights, and measures proposed for the United States of America* championed decimals and the ease, as Bordley put it, of 'dividing by dots'. Supportive letters appeared in New York and Philadelphia newspapers, and universities like Columbia and Pennsylvania lobbied for reform. Scanning the public and political landscape, the historian Julian Boyd, editor of Jefferson's *Writings*, estimated that 'if ever a moment existed in which the public mind seemed ripe for a general reformation and in which political circumstances seemed auspicious, the summer of 1790 was assuredly that moment'.

For the converts it had become a race to implement the new scientific system of measurement. Once a pattern of measurement and disposal of the western lands was established, it would grow as resistant to change as the rest of the country. 'Too long a postponement,' Jefferson warned, '. . . would increase the difficulties of it's reception, with the increase of our population.'

How short a time they had could be deduced from the pressure that Rufus Putnam was putting on the federal government to extend its power westward. In late 1790, Putnam wrote to Hamilton urging the United States to take over the Spanish-controlled Mississippi, otherwise the Northwestern Territory could not flourish. His frustration was born from the problems of the Ohio Company. That winter his wife Persis joined him with the rest of the family, including three unmarried daughters and two grandchildren, but they were still living inside Campus Martius. The settlement of Marietta was growing fast, a school had been established, and some settlers had purchased lots far upstream on the Muskingum, pushing towards the flatter country that lay beyond the Alleghenies. But to become profitable, the Company needed a surge of settlement, and it was the difficulty of shipping timber and produce down the Mississippi that most deterred would-be customers. With the United States in control of the river, the last bar to settlement west of the mountains would be removed, and the population of the Ohio valley would explode.

Despite living inside the fort, Rufus had no real fear of the territory's native occupants. Treaties had been signed at Fort Stanwix and Fort Harmar with the Six Nations and some of the Western Confederacy whose territory stretched from Michigan to the Ohio river. It was, however, recognised that care had to be taken. The surveyors who were the first into the territory had always suffered from having their horses stolen by Indians, and individuals travelling alone were recognised to be at risk; but there was no immediate apprehension of greater danger. Then in the winter of 1790 a column of troops led by General Harmar was ambushed by Indians, and in January 1791 news reached the fort that fourteen people had been killed in one of the Muskingum settlements. Rufus communicated at once directly with President Washington.

'Our prospects are much changed,' he wrote in his wayward hand. 'In stead of peace and friendship with our Indian neighbours, a hored Savage war Stairs us in the face the Indians appear ditermined on a general War.'

Of the dozen or more nations in the Western Confederacy, those most affected by the Ohio settlements were the Delaware, Wyandot and Miami. Although they hunted further west in the winter months, in the spring and summer they cleared trees to plant corn, beans and squash in the region around the foothills of the Alleghenies, into which Israel Ludlow and his surveyors were increasingly moving. Their apprehension at the straight lines being scythed through the forest, and the growing number of square farms springing up in their wake, now spilled over into warfare.

Responding to Putnam's plea for help, Washington ordered General Arthur St Clair to take an army of two thousand men to punish the Western Confederacy. With fatal over-confidence,

St Clair led his men up the Miami river from Fort Washington near Cincinnati, and on 4 November blundered into an ambush set by the Miami war chief, Little Turtle. In the bloodiest defeat the United States was ever to suffer in its Indian campaigns, 630 men were killed, and another 270 wounded.

St Clair's force was handicapped by defective weaponry and transport that had been supplied by William Duer and almost certainly originated with a company owned by his fellow speculator and fraudster, Alexander Macomb. Even before this was known, many people understood that the United States Army was fighting native inhabitants in order to protect land companies' profits. New England newspapers in particular were scathing. 'Let offensive operations cease,' the *Connecticut Courant* roundly declared. 'They are calculated for land jobbers only.' When three hundred volunteers were sent to reinforce General Harmar, the *Boston Gazette* claimed scornfully that they were really investors 'known by the Name of Paper-Hunters or Hamilton's *Rangers*'.

To buy time while another army was assembled under General Anthony Wayne, the Secretary of War, Henry Knox, instructed Putnam to arrange a meeting with the Western Confederacy. 'You will make it clearly understood that we want not a foot of their land,' Knox ordered, none too honestly, 'and that it is theirs and theirs only – That they have the right to sell and the right to refuse to sell.'

Rufus did his best to follow these instructions at a council held in Vincennes in the summer of 1792, but he was never good at dissimulation. The Miamis and most of their allies stayed away, and those who did attend clearly did not believe him. 'I would have been glad if matters had remained as they were in the days of the French,' said the chief of the Kaskaskias, Jean-Baptiste Ducoigne. 'Then all the country was clear and open. The French, English and Spaniards never took any lands from us. We expect the same of you.' The chief of the Potawatomies recalled that he had refused a British attempt to buy his land, saying, 'I foresaw that if I parted with my land I should reduce the Women and children to weeping.' Each speaker made the same point, that the United States should advance no further west. 'It is best that the white people live in their own country and we in ours,' said Ducoigne. 'We desire of you to remain on the other side of the river Ohio.'

Since there was no chance of that happening, the inhabitants of the Northwestern Territory had to be made to change their minds. Only one body could do that. The 1787 Land Ordinance expressly reserved to the United States government the right to acquire land from the Indians. But until Wayne's army was in a position to persuade them, further settlement west of the Ohio ceased.

The pause had two effects. For several years settlers poured south rather then west. Rather less expectedly it gave more time for the reform of the United States' weights and measures.

NINE

Annie's Ounce

THE FATE OF JEFFERSON'S PROJECT lay with the Congress, and at the first opportunity in December 1790, when Congress reconvened in Philadelphia, a committee was appointed to consider the Secretary of State's proposals. Its membership of five included James Monroe, a supporter, but also the hostile Robert Morris, and the report published in March 1791 reflected the latter's views. It noted that both France and Britain were considering how to obtain 'a uniformity in the measures and weights of the commercial nations', and that as this 'would be desirable', the United States should wait to see what the others would do.

The Senate might be unwilling to take the lead, but the President now swung his enormous prestige behind the object. On 25 October 1791 Washington returned to the need for reform, explicitly emphasising the advantage of putting it on the scientific basis that Jefferson had argued for in his report. 'A uniformity in the weights and measures of the country is among the important measures submitted to you by the Constitution,' Washington declared in his annual address to Congress, 'and, if it can be derived from a standard at once invariable and universal, must be no less honorable to the public councils than conducive to the public convenience.'

It was impossible to ignore such a command, and less than

a week later the Senate appointed a three-man committee led by Ralph Izard to report on the subject. This time there was no Morris among its members, but Monroe was still there. When it reported in April 1792, it was unanimous in recommending Jefferson's second scheme – for complete reform. The committee requested the President to have a second's rod constructed at public expense, adding, 'That the standard rod so to be provided shall be divided into five equal parts, one of which, to be called a foot, shall be the unit of measures of length for the United States. That the foot shall be divided into 10 inches, the inch into 10 lines, the line into 10 points; and that 10 feet shall make a decad, 10 decads a rood, 10 roods a furlong, and 10 furlongs a mile.'

Everything Jefferson had wanted was there: 'That measures of surface in the United States be made by squares of the measures of length; and that in the case of lands, the unit shall be a square, whereof every side shall be 100 feet, to be called a rood.' The basic measure of capacity was a cubic decimal foot, to be called a bushel, and the basic unit of weight, the ounce, was to be derived as Jefferson had suggested from the weight of a cubic decimal inch of water.

With the benefit of hindsight it is clear that the moment was critical. Ten millennia of traditional, human-based weights and measures were approaching their end, and for a short period everything was in flux. Since that moment, science and technology have required ever more exacting standards of accuracy in measurement. In concept, either of the two scientific, decimal systems put forward during that period – France's* and Jefferson's – could have met their needs. Today one of those systems dominates the world, and it might have been Jefferson's. No scientific project of similar consequence was ever so utterly dependent on the vagaries of fate and human nature.

* See Chapter Ten.

The strongest opposition to any change came from the merchants most closely tied to trade with Britain; but even among that block of opinion there was a growing awareness of the need for more certainty in weights and measures than the traditional system could offer. To meet France's apparently insatiable hunger for wheat, for example, American exporters needed to buy from farmers outside the traditional growing areas in Pennsylvania and New York state. Each summer in the 1790s, the disputes that arose over payment for cargoes of grain brought home more forcibly than any political argument the disadvantage of having no agreed size of bushel. Simply in the interest of efficiency, a great tycoon like the diminutive John Swanwick, who had been Robert Morris's partner before creating his own business empire, was now prepared to accept that some certainty had to be introduced into the system.

Nevertheless, what the reformers' cause required above all was international support for change, and especially from one of the United States's major trading partners, either its old ally France, or Britain, rapidly growing to be the dominant economic and industrial power of the age. It was for this reason that politicians in Philadelphia, and as it happened in Paris, read with such interest the words of an otherwise obscure English parliamentarian named Sir John Riggs Miller. In July 1789, before either Jefferson or Talleyrand had started work, Riggs Miller had made a long-winded speech to the House of Commons criticising the deplorable state of British measures and calling for 'one general standard of weights and measures to be observed throughout the kingdom'.

Riggs Miller was a bore – the evidence lies in his interminably dull speeches – but he was also persistent and driven by genuine outrage at what he called 'abuses fertile in confusion to commerce, and in distress and difficulty to the poor'. He kept the House of Commons sitting late with his accounts of the scandalous discoveries made by the Carysfort Commit-

The young George Washington holding the surveyor's essential tools.
In his left hand, Gunter's chain, and in his right a circumferentor or
primitive theodolite.

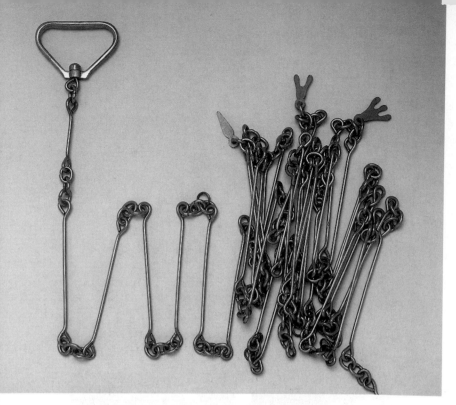

Above Gunter's twenty-two-yard, hundred-link chain, whose length can be found in a cricket pitch and in the dimensions of every city block in the United States.

Right Gunter's quadrant or quarter circle, for use in navigation and astronomy.

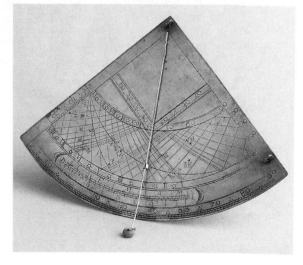

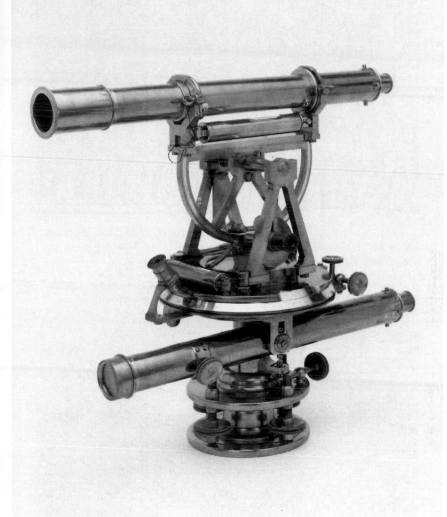

A superb eighteenth-century theodolite made in 1737 by J. Sisson, one of the finest of London's gifted instrument manufacturers.

Sixteen and a half feet make a rod, pole or perch (the last half foot is just visible on the left). A German representation in 1583 of the basic land measure as it was becoming standardised.

A sixteenth-century scene of two surveyors using cross-staffs to measure distance. The ends of the cross-bar on each staff are aligned with the end-points of the length to be measured.

Thomas Jefferson (1743–1826). Third President, author of the Declaration of Independence, and instigator of the squares that spread across the nation.

Connecticut settlers entering the Western Reserve with Conestoga wagons and the expectation that Moses Cleaveland's survey teams have laid out the land for them.

Rufus Putnam (1738–1824). The only known image is unexpectedly benign.

The Rev. Manasseh Cutler (1742–1823). Serpentine, subtle and 'given to making himself agreeable'.

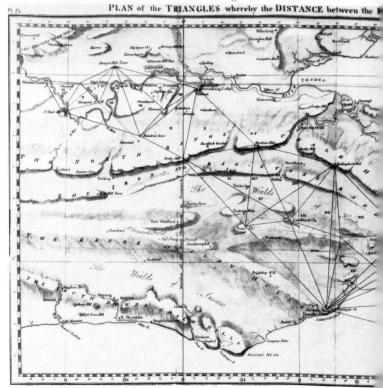

The Paris–Greenwich triangulation. The point where the Cassinis' triangles met William Roy's. The first base-line is measured with minute care, and a triangle drawn, whose sides become the base for further triangles.

William A. Burt (1792–1858), Michigan surveyor and inventor of the solar compass.

Ferdinand Rudolph Hassler (1770–1843). The United States's master measurer.

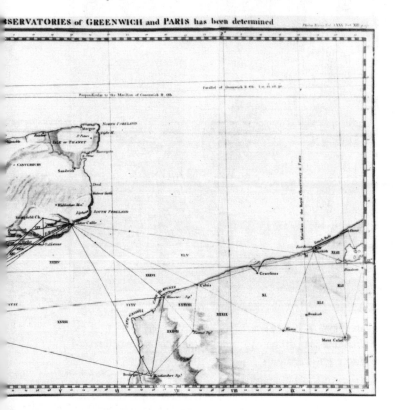

Usage des Nouvelles Mesures.

1. le Litre (*Pour la* Pinte)
2. le Gramme (*Pour la* Livre)
3. le Mètre (*Pour l'*Aune)
4. l'Are (*Pour la* Toise)
5. le Franc (*Pour une* Livre Tournois)
6. le Stere (*Pour la* Denue Voie de Bois)

J.P. Deliou G..... inv.

Labrousse Sculp.

A Paris chez Deliou Rue Montmartre N° 142. pres le Bouleo

A poster dated about 1800 attempting to reconcile the French to the joys of metrication in, respectively, capacity, weight, length, area, currency and quantity.

tee thirty years earlier, which were to provide Jefferson with such good material. The Act of 1700 alone, the committee reported, managed to define the Winchester bushel, used on both sides of the Atlantic, in four contradictory ways.

In the circumstances, it was hardly surprising that the committee had been unable to find a single bushel container or a single weight in the office of the Exchequer, or the Tower, or the Mint, that was the same size as any of the others. But the copies made from these standards for use in cities throughout Britain were worse still: 'The Unskilfulness of the Artificers [makers], joined to the Ignorance of those who were to size and check the Weights and Measures in use, occasioned a great Number of different standards to be dispersed throughout the kingdom which were all deemed legal yet disagreed [with each other],' the committee reported. Poorly-made copies of the bushel and the yard were sent out to cities and 'became themselves again the Standards for other Copies, made and used in different [places] by Artificers yet less skilful than the first Makers . . . Thus every Error was multiplied till the Variety . . . rendered it difficult to know what was the Standard and impossible to apply any adequate remedy.'

At Riggs Miller's request, the House of Commons ordered all market towns to report what measures they used, and in February 1790, when he addressed the House again, he had enough evidence to show that the situation was if anything worse than in 1758. Given his record as an orator, the House of Commons was probably empty by the time Riggs Miller reached his point, but it would have been worth waiting for. 'Nothing,' he thundered, 'could be more degrading, more absurd, and more preposterous than that that which should be the clearest and plainest . . . should prove to be undefined in law, obscure in practice and involved in perplexity.'

He was undoubtedly right, but in most circumstances there was no great need for precision. For all the confusion, it was possible for Annie Watt to write in the 1780s from Glasgow

to Birmingham, where her husband was overseeing the construction of his steam-engines:

> Dear Jamie, If you could get the under written
> articles without much trouble the purchases would
> be better than we could get them here:
> 3 lb of Jordan almonds
> 1 lb of Black Pepper Corns
> Half lb of white Ginger
> Quarter lb of White Pepper Corns
> 1 oz of Nutmegs
> 1 oz of Cloves
> 1 oz of Maize
> 1 oz of Cinnamon
> If a small Barell of Anchovies could be got fresh
> they are very useful. We buy them here from 1/6
> to 2/ per lb. Could you get us some Garden Seeds
> if I was to send a list –.

No doubt the Birmingham pound was different from the
Glasgow one, but the fact that Mrs Watt could specify ounce
quantities for what sounds like a recipe for mulled wine suggests that the difference did not matter greatly. Cookbooks
still offered recipes without quantities, or with the advice to
use 'a handful of flour' or 'a dash of pepper', and even when
amounts began to be specified, as in Mistress Margaret Dods'
1829 *Cook and Housewife's Manual*, they do not sound very
exact: 'To every pound of fresh, juicy, rump beef, allow a quart
of soft water,' run the instructions for her stock-broth, 'and
to this add any fresh trimming of lean mutton, veal, poultry
or game which the larder affords.' It is not one of her most
mouthwatering recipes, and since old fowls, rabbits, onions,
turnips and herbs might also be tossed in and boiled up,
precision was clearly not important.

Unlike France, there is no evidence that in Britain the old
local measures were undergoing wholesale change. As a result,

the injustice they entailed – 'The poor thresh out corn and other grain by the largest customary bushel and buy their bread by the smallest,' declared Riggs Miller – did not arouse the same resentment, simply because this was the way it had always been for the poor.

Yet Riggs Miller's attack on the old measures did reflect a need for uniformity that was appearing in other areas, especially trade. Earlier in the century, Glasgow tobacco-merchants had found it impossible to cope with the different sizes of hogshead used by Georgia, the Carolinas and Virginia for measuring tobacco, and insisted on a standard size based on Virginia's. The Treasury did the same with the dimensions of liquid containers after it was caught out by Thomas Barker, a wine merchant who imported Spanish wine in beer butts holding 150 gallons but only had to pay duty on 126 gallons, because according to Treasury regulations that was all a wine butt was supposed to hold.

The growing Industrial Revolution also needed precise, standard units. The importance of precision was illustrated by William Small when he returned to Britain after leaving the United States in 1764 and, as a friend of both men, brought together the industrialist Matthew Boulton and the inventor James Watt. His own contribution to that partnership, which served as the catalyst for the Industrial Revolution, was to design micrometers capable of measuring distances as small as one-hundredth of an inch for Watt, so that the valves needed by the separate condenser in his improved steam-engine could operate efficiently. Standard units were also essential for the Carron Company ironworks outside Edinburgh after it won a contract in 1778 to supply the Royal Navy with thousands of cannon. Although Scotland's measures were officially made identical with England's in 1707 by the Treaty of Union between the two countries, in practice its longer mile, its larger acre and a host of units like the stoup and the mutchkin remained in common use for another two centuries.

But where the calibre, weight and dimensions of the new guns were concerned, using Scots units would have been disastrous, because the Carron artillery had to match the powder, ammunition and gun-carriages produced in other parts of Britain.

It was the scientists who needed accuracy more than anyone, and in the absence of government standards they provided their own. In 1742, as Jefferson noted, the Royal Society had a yard standard built to very exact specifications based on Elizabeth I's brass yardstick, and the Carysfort Committee commissioned a superb example from John Bird, the foremost instrument-maker in London. His yard, made in 1760, was a work of art, a thirty-nine-inch brass bar with the two hairline grooves exactly thirty-six inches apart – as gauged against the Exchequer yardstick of 1601 – set into gold studs in order to avoid distorting the bar itself. As a secondary check in case it was destroyed, Bird's bar was measured against a second's pendulum oscillating in a vacuum. Eventually it would become the master yardstick against which the standards of Britain and its empire and the United States would be measured.

The usefulness of that scientific insistence on exactness first became obvious to a wide audience in 1785 – thanks to the French mania for measuring the size of France. When the most northerly of the surveyors' triangles were about to reach Dunkirk, Cassini de Thury suggested to the Royal Society in London that they should be continued across the Channel. This would enable the distance between the observatories in Paris and Greenwich to be precisely measured so that their observations could be co-ordinated.

The Royal Society, founded in 1663, was financed by people interested in science, unlike the Académie, which was government-funded. It had the advantage of flexibility and innovation in the encouragement of science, but lacked the resources for the sort of strategic approach that the Académie had taken to geodesy. Consequently in Britain, while numerous county maps existed, and the Highlands of Scotland had been systematically surveyed for military reasons following the Jacobite uprising of 1745, there was nothing like the national map based on exact triangulated measurement that France now possessed. Nor was there much enthusiasm for helping the French.

However, General William Roy, the formidable Scots disciplinarian who had masterminded the Highland survey, had other ideas. Linking the observatories had no particular interest for him, but it did offer an opportunity to fulfil his lifelong ambition of creating a British map to rival the French. His advocacy helped persuade the Royal Society to back the project, and with the enthusiastic attendance of King George III and much of London's fashionable society, Roy began to measure out a five-mile base-line at Hounslow Heath, where Heathrow airport is now located. Flat ground is as welcome to surveyors as it is to commercial aircraft, and once the undergrowth that made the area a haven for highway robbers had been cleared, there should have been nothing to hold up work. The one essential quality of all great surveyors, however, is an infinite desire for greater accuracy, and Roy passed that test easily.

To calculate their base-lines, the French had used wooden measuring rods which were protected in boxes and checked daily against an iron bar exactly one *toise* in length. These were tried but discarded by Roy, because even under cover the wood tended to expand slightly in a damp atmosphere. He replaced them with calibrated, eighteen-foot-long glass rods, manufactured by the great instrument-maker Jesse Ramsden. They needed to be carried in special containers and supported on trestles to prevent them from fracturing, and were checked against the Royal Society's brass yard, which had been marked out with a beam compass fine enough to show subdivisions of one eight hundredth part of an inch.

But nothing shows Roy's fanaticism to better advantage than his choice of a theodolite. Jean-Charles Borda, the French cartographer in charge of the final stage of triangulation, had invented a beautiful instrument that carried observation to a new order of accuracy. It had long been recognised that one sight on a distant mark through a theodolite's narrow focus might contain an error, and geodesists as a matter of habit would take five or six observations and average the results. Borda's genius was to build into an instrument weighing no more than twenty pounds a mechanism that automatically averaged ten observations, and thereby reduced the margin of error from fifteen seconds' deviation to little more than one second.

Borda's repeating theodolite was the pride of French geodesy, but when it was offered to Roy, he turned it down. It had one major drawback: for maximum accuracy the mechanism had to be adjusted constantly. But that was not Roy's objection – he wanted more than its maximum. For that he went back to Jesse Ramsden and commissioned the most precise instrument that had yet been constructed in surveying history. The thing was a monster, a mass of gleaming brass weighing in at over two hundred pounds, with a three-foot-diameter circle and an achievable accuracy on its giant compass of less than one second. To move it Roy needed a company of redcoats on hand at all times, but it offered measurements as exact as the eighteenth century could aspire to. While he lived, it soothed the search for perfection in the General's Calvinist soul.

Such exquisite engineering takes time, and it was five years before the line of triangles reached from Greenwich to Dover, from where sightings across the twenty-two-mile stretch of the English Channel connected the British survey to the French. The accuracy of the work, however, could not be questioned – triangulation showed that the distance between the observatories in Greenwich and Paris was just fifteen inches shorter than that calculated by astronomical observation. Roy, unfortunately, had little time left to apply his ruthless zeal to the larger task of measuring Britain before he died in 1790, and those who came after him lacked his burning flame.

The project was taken up by the army's Board of Ordnance, which had responsibility for artillery, and in case of invasion would need accurate knowledge of the range from one position to another. In competent, military fashion it produced the first triangulated map of the United Kingdom in 1822, but compared to the standards set by the French and Roy the result was rushed and filled with errors. Nevertheless, it triggered a continuous cycle of improvement, leading eventually to the complete triangulation of the country in 1853, on a scale of six inches to the mile. The zeal for mapping that was unleashed by the Ordnance Survey had consequences beyond Britain, for it led to triangulated maps being made of almost every square mile of the Empire, from the immensity of India to the compactness of Cyprus.

It was the lack of a personality like Roy's that doomed the British effort to spread exactness from mapping into weights and measures in general. Riggs Miller understood the need, but he lacked the dynamism. In April 1790 he initiated a debate in Parliament calling for the wholesale reform of the system. He suggested four possible bases for a new scientifically based set of decimal measures: the weight of a drop of water, the distance covered by a falling body in one second, the distance covered by a degree of a meridian at the equator, and finally the length of a second's pendulum. This last was his personal choice, although as a patriotic Briton he specified that 'the pendulum which vibrates seconds at London, is the most proper standard for Great Britain, and a medium for all Europe'. This length would give the basic unit of capacity, and weights would be derived from the weight of rainwater in a container of that capacity.

During his speech Riggs Miller quoted from a letter written to him by Talleyrand to show that the co-operation between France and Britain over the survey might be extended to reform of measurement. 'I understand that you have submitted for the consideration of the British Parliament, a valuable plan for the equalisation of measures,' wrote Talleyrand. 'I have felt it my duty to make a like proposition to our National Assembly. It appears to me worthy of the present epoch that the two Nations should unite in their endeavour to establish an invariable measure and that they should address themselves to Nature for this important discovery. Too long have Great Britain and France been at variance with each other, for empty honour or for guilty interests. It is time that two free Nations should unite their exertions for the promotion of a discovery that must be useful to mankind.'

'A turd in a silk stocking' was how Napoleon described the smoothly sinister Talleyrand, and probably any MP still awake at the end of Riggs Miller's interminable speech would have yawned his agreement. Nevertheless, the House agreed to appoint a parliamentary committee to examine his ideas, and so in the summer of 1790 London, Paris and New York were all considering the same solution – decimal units based on the second's pendulum – to the same age-old problem of local, variable measures. Such complete synchronicity might have moved mountains – even shifted public opinion.

TEN

The Birth of the Metric System

O N 19 JUNE 1791 a detachment of the Paris National Guard posted outside the royal apartments in the Tuileries palace admitted a group of twelve men to see the King. It is possible that the soldiers recognised two of them, citizens Marie-Jean Condorcet and Gaspard Monge. Both were prominent members of the National Assembly, but the names of others, notably Joseph-Louis Lagrange, now famous for his work on mechanics and differential calculus, Pierre Laplace, renowned as the founder of celestial mathematics and supreme exponent of probability theory, and Antoine Lavoisier, discoverer of the properties of oxygen, are probably better-known today, at least among mathematicians and chemists, than they would have been by the royal guard.

That Louis XVI had commanded their presence in the palace on that day throws light on a pleasing aspect of his otherwise rather dim personality. He was neither imaginative nor energetic, but he was interested in how things worked. His guests on this occasion were the members of the Académie des Sciences's commission on weights and measures which would establish the basis of France's new decimalised system of measurement. The depth of the King's interest can be inferred from the date of their visit. Alarmed by the growing power of the Revolution, and terrified that the Paris mob that prowled around the streets outside the palace would break in and lynch him and his family, Louis had made plans to escape that very night.

Shortly after the scientists had left, the King wrote a long justification of the action he was about to take, then secretly changed into a servant's round hat and plain coat, while the Queen, Marie-Antoinette, disguised herself as a governess. At midnight on 20 June Louis slipped past the guards, followed some minutes later by the Queen, and together they joined their children in a covered coach. The escape attempt almost succeeded. Despite the slowness of the coach, and an accident crossing a narrow bridge, they reached Varennes, barely three hours from the border, before a Revolutionary guard compared the servant's face with the image of the King on a banknote and recognised him. Louis and his family were brought back to Paris as prisoners, and he spent the rest of his life more or less under house arrest. Less than eighteen months later he was guillotined. Thus virtually his last autonomous act as a free man was to discover how the new decimalised system of weights and measures proposed for his kingdom actually worked.

Following Talleyrand's recommendation for reform in March 1790, the Assembly had ordered the Académie to set up a commission on the subject. Its first report appeared in October, recommending that the system to replace the old aristocratic measures should be decimal. The arguments were those that Jefferson had marshalled: the simplicity of calculating in tens, and the opportunity that this gave the average citizen of dealing on equal terms with those more educated than himself. Soon afterwards, a new member, Gaspard Monge, an ardent Jacobin but also a mathematical prodigy who is now considered the father of differential geometry, replaced Tillet on the Académie's commission. That winter it turned its attention to the choice of a scientific basis for the new system.

There were only three possibilities. The commission's

report, published on 19 March 1791, listed them: 'the length of the pendulum, a quadrant of the circle of the equator, finally a quadrant of the earth's meridian'. As scientists, they clearly had to examine each possibility, but the report made apparent the clear advantages of the second's pendulum. The length was simple to calculate, and the results easily checked, so that the experiment could be run anywhere in the world without the need to go back to the place where the original trial was made. 'In fact, the laws [of physics] concerning the length of the pendulum are sufficiently certain, sufficiently confirmed by experience to be used in experiments without fear of any but imperceptible errors.' None of this was new, but the endorsement of such a distinguished panel was impressive confirmation of the method's reliability.

For the first eight pages of the French commission's report, the unity between the reformers in three different countries was maintained. Then on page nine, they sprang a bombshell. Having made the case for the second's pendulum, they suddenly commented: 'However, we ought to observe that the unit thus derived contains something arbitrary. The second of time constitutes one eighty-six thousandth part of the day, and is consequently an arbitrary division of that natural entity. Thus, to fix the unit of length, requires not only a heterogeneous element – time – but one that is arbitrary.' The report acknowledged that if necessary it would be simple to gauge the pendulum's length against the duration of a day rather than a second, thus retaining the 'natural entity' undivided (a pendulum that ticked a hundred thousand times in a twenty-four-hour day would measure twenty-seven inches), but this alternative was rejected too on the ground that it used 'a heterogeneous element'.

Having set aside the one internationally accepted foundation for a scientific system, the commission stated their preference for a unit taken from the earth itself, because – and the argument is hardly overwhelming – this would be 'analogous to all the real measures which in everyday life are also taken on the earth'. They then swiftly dismissed the possibility of using the equator as a basis – too far away, too expensive, too complicated – leaving the meridian as the one remaining option. 'In the end,' they concluded almost frivolously, 'one could say that everyone lives on a meridian, but only a part [of humanity] lives on the equator.' At which point the report baldly declared, 'A quadrant of the earth's meridian will therefore become the [basis] of the measure, and the ten millionth part of that length will be the unit used.'

This then was to be the basis of the metric system, and it could hardly have been chosen in more capricious fashion. Today the length of the metre is defined in terms of time – it is the distance travelled by light in a vacuum in a fraction of a second – and the fact that such a definition is, in the commission's terminology, 'heterogeneous' is irrelevant. The clinching argument for choosing a quadrant of the meridian, from the equator to the North Pole, was no more logical: 'In fact,' the commission asserted, 'it is a lot more natural to record the distance from one place to another in terms of a quadrant of one of the earth's great circles than to record it in terms of the length of a pendulum.' Five of the finest scientific minds in the world, each trained in the merciless school of Cartesian dialectics, ought to have been able to come up with something more persuasive.

The speciousness of the commission's reasoning was immediately apparent to contemporaries, and inevitably triggered speculation about the true reasons for their choice. The first clue came from the curious meeting of the scientists with Louis XVI in June 1791. It was the King, as the executive power, who signed the Assembly's decree, and who commissioned the Académie to appoint the appropriate people to carry out the required experiments; but it is unlikely that, amidst the collapse of the *ancien régime* and the howls of the mob, the reform of weights and measures occupied many of

Louis XVI's waking thoughts. One aspect of it, however, clearly snagged his curiosity.

Having chosen the meridian as its basis, the commission recognised that its distance would have to be measured accurately. Although it was not necessary to survey all ninety degrees from equator to pole to produce a reliable estimate, the longer the extent of it that could be measured, the more accurate the estimate; and to reduce errors to the minimum both starting and finishing points needed to be at sea-level. The line that suited all these requirements extended from Dunkirk through Perpignan to Barcelona. The length measured would cover slightly more than nine and a half degrees, enough to extrapolate with a good degree of certainty the full distance. French geodesists had already triangulated most of it twice. For the new metre, the commission recommended that it be triangulated for a third time.

Portly, dutiful and burdened with a wife considerably brighter and livelier than he, Louis had not lost his enthusiasm for asking why and how things worked. Regardless of his imminent plan to escape, he summoned the scientists to the Tuileries. When they were assembled, he turned to Jacques Cassini, the fourth of the family, who had been put in charge of measuring the meridian, and asked why he was going to repeat a measurement which his father and grandfather had already done before him. 'Do you think you can do it better than they?' the King demanded.

Concealing the fact that he had already shuffled off responsibility for this physically demanding task onto two other scientists, Pierre Méchain and Jean-Baptiste Delambre, Cassini offered this answer: 'Sire, I only flatter myself that I can do better because I have a great advantage over them. The instruments they had only measured angles to an accuracy of fifteen [degree] seconds. Monsieur le Chevalier de Borda has invented one which measures angles to an accuracy of one second. That's my entire justification.'

The answer revealed that a shift of power had taken place, away from the idealistic, international dreams of Condorcet and towards the pragmatic ambitions of Jean-Charles Borda, the commission's chairman. Borda wanted the meridian to be used as a basis because it would have to be measured with the superb repeating theodolite that he had invented, and the credit for completing the work begun by Picard, continued by generations of Cassinis, and involving expeditions to Peru and Lapland, would finally go to him. Consequently he refused to accept the impossibility of measuring the exact distance from the equator to the pole. In his *Essai sur l'histoire générale des sciences*, published in 1803, the physicist Jean-Baptiste Biot also suggested that this was why the pendulum had been abandoned. Borda, he said, had persuaded the commission that the meridian had to be selected because it would enable them to achieve the goal of French science for more than a century, that of establishing the size of the earth.

By itself that argument might not have been sufficient to persuade the commission to abandon the second's pendulum, but it was backed by another still more compelling motive, one familiar to generations of scientists and funding bodies since then: establishing the length of the meridian was a bigger, more expensive research project than establishing the length of the second's pendulum. On these grounds their choice was shrewd, for the National Assembly did indeed appropriate 300,000 *livres* for the project, a sum that kept many of the Académie's scientists in work when its annual ninety-three thousand *livres* grant was cut in 1793.

Although Borda's voice was decisive, the influence of Monge, politically naive and a decimal extremist, should not be overlooked. In 1793 he became the driving force in the committee responsible for decimalising the number of months in the year and of days in the week, and for proposals to decimalise the number of hours in the day and, naturally, of seconds in the minute. Clearly, if the committee began by basing measurement of distance on a pendulum ticking to the old second, all their work would have to be torn up when they came to decimalise time. Monge, the Jacobin, had his own radical agenda. For him the meridian metre was only the first step in a world made up entirely of tens.

There was a certain irony in this outcome, that it should have been Talleyrand, the most devious and dislikeable of politicians, who gave expression to the high ideals of science, while the scientists abandoned their principles for materialist, nationalist and political considerations. Nevertheless, on 26 March 1791 the National Assembly accepted the report of the Académie's commission in its entirety, and in a formal decree made the first political break with the long history of organic measurement.

Once the meridian's distance was established, one ten millionth part of it would become the standard measure, and from it would be derived a decimal system of lengths. At the same time, the decree ordered that two other experiments should be made: the first to establish the number of oscillations a pendulum of that basic length would make in a day (this would serve as the means of verifying the results elsewhere in the world), and the second to establish the weight of a cubic measure of water. This would provide the basis for the new decimal system of weights.

Thus what might have been an international enterprise became a French project. In London, the commission's choice tended to confirm ancient suspicions about France. Riggs Miller had failed to win re-election to the House of Commons, and in his absence the impetus for reform was fading. The government had already let it be known that it regarded the French commission's proposals for decimalisation as 'impracticable'. Now it effectively turned its back on proposals so closely associated with a country which had always been Britain's rival, and with which from 1793 it was at war.

The most severe impact was felt in the United States, where the French commission's report cut the ground from beneath the feet of Thomas Jefferson. He had specifically tied his proposal to French science, deliberately abandoning measurement of the earth as a basis for his system, and going so far as to choose the latitude of Paris as the point where the definitive measurement of the second's pendulum should be made. 'The element of measure adopted by the National Assembly,' he now wrote in despair, 'excludes, *ipso facto*, every nation on earth from a communion of measurement with them.'

Despite the unanimous endorsement of his scheme by Senator Ralph Izard's committee, neither Jefferson nor Monroe could line up a Senate majority in favour of its recommendations. The President continued to press for reform, and enthusiasts for change could be found on each side of the divide that began to open up in 1791 as the legislature divided into two political groupings, Jefferson's Republicans and Hamilton's Federalists. But the evil consequences of Borda's last-minute switch hampered all their efforts. The existence of two proposals for reform encouraged more. The New Yorker Robert Livingston, later to be Jefferson's Minister in France, developed his own scheme, and Oliver Wolcott, a Federalist who was to be Secretary of the Treasury under John Adams, wanted another version similar to Jefferson's first, moderate suggestion.

It was soon apparent that the problem that concerned the Senate was not whether change should be made, but what form it should take. By the spring of 1793, it had considered the Izard report three times, and recommended three different courses of action, but on each occasion had accepted that the United States's weights and measures needed to have a scientific basis, and that the most convenient basis was the second's rod. After the latest debate ended in deadlock, the exhausted senators voted to postpone further consideration of the subject, and turned their minds to other matters.

Jefferson retired as Secretary of State at the end of 1793, frustrated by the gathering power of the Federalists – and thereby earned Rufus Putnam's undying hostility for deserting the President. But he never gave up his belief in decimals. As President from 1801 to 1809 he would at last succeed in decimalising a second scheme to measure the United States, and in his *Autobiography*, composed in his seventies, he wrote with undiminished optimism, 'The division [of the dollar] into dismes, cents and mills is now so well understood, that it would be easy of introduction into the kindred branches of weights and measures.'

Nor did the campaign for reform end with Jefferson's departure. President Washington continued to support it, the Senate accepted the principle of change and, as would soon become evident, opinion in the House of Representatives was also in favour. Crucially, out on the frontier the window of opportunity remained open. Through 1792 and 1793, General Wayne's offensive against the Western Confederacy proceeded so cautiously that it was hardly apparent. During those years he constructed a string of forts to the west of the Indians' main force, so that they were gradually encircled. But so long as the Confederacy's soldiers remained undefeated in the Ohio forests, no settlement of the Northwestern Territory could take place.

While the military situation still hung in the balance, a fresh impetus for change came from an unexpected direction. In France a new government took power in September 1792, when the former combination of King and Assembly was set aside in favour of a purely republican constitution. The temper of this new government – the Constituent Convention – was apparent in its decision to guillotine Louis XVI in January 1793 and, more importantly in the long run, to conscript 300,000 young men to fight in the army. Confronted by war on its frontiers and rebellion at home, the government set up a small, nine-member Committee of Public Safety with the power to co-ordinate every activity relevant to the country's security. Among these were the researches of the commission on weights and measures. The results of their work were now needed urgently. The old weights and measures were still in use, but because so many of the standards had been physically removed, uncertainty was giving rise to increasingly ugly disputes in the marketplace and at every rent-day.

Under pressure, the Académie's commission produced a third report in the spring of 1793 recommending that the new unit of length be named a '*mètre*'. Using the results obtained by La Caille in 1740, which showed that one degree on the meridian measured 57,072 *toises*, the commission estimated provisionally that the metre's length would be 443.44 lines, or in British terms, slightly more than thirty-nine inches. Monge's growing influence was apparent in the commission's other recommendations that time, geometry and the calendar should all be decimalised as well.

On 1 August 1793 the Convention accepted the main points of the report – length to be based on the metre, area on the *are*, or a hundred hundred square metres, and capacity on a cubic *decimètre* (one-tenth of a *mètre*). The name for this last unit was to be a *pinte*, and the basic unit of weight, named a *grave*, would weigh as much as a *pinte* of water.

All this the scientists welcomed, but there was a sting in the tail. The new system, the Convention decreed, was to be in place twelve months later, and on 11 September a temporary committee of weights and measures, made up of the old commission, was created to put the decree into effect. It was an impossible timetable. The survey of the meridian from Dunkirk to Barcelona was less than a third complete – starting from the north, Delambre had barely reached the Loire, while Méchain, who had set out from Barcelona, was only just beyond the Pyrenees – and years would be needed to finish the work and to check the results. If the survey had to be abandoned, so too would be Borda's cherished project of establishing accurately the size of the earth. Under this threat, splits began to appear in the temporary committee.

Borda and most of the committee's members protested vigorously. Lavoisier in particular warned that, in adopting a slimmed-down speedy solution, they risked losing 'a general system, which embraces geography, navigational skills, surveying, weights, currency and measurements of solids and liquids. In a word, it would mean losing, perhaps forever, the inestimable advantage of eliminating all calculation problems by the use of decimal divisions.'

This showed courage, for the Convention was beginning to focus on the Farmers-General, the tax organisation of which Lavoisier had been a member, on the grounds that it had made excessive profits and squeezed taxpayers unjustly. In November, the all-powerful Committee of Public Safety ordered Lavoisier's arrest. Again Borda protested. 'The presence of citizen Lavoisier is irreplaceable,' he explained to the Committee, 'because of his unique talent for anything requiring the utmost precision. It is urgent that this citizen should be able to carry out the work he has always performed with as much zeal as energy.'

Monge, however, remained silent, and his reward came that winter when the Committee of Public Safety suddenly lost patience with the temporary committee and declared that 'responsibility should only be delegated to those who showed they were worthy of it by their republican virtue and their detestation of kings'. Borda, Lavoisier, Laplace, Delambre and others were dismissed, but Monge remained, the one real giant, however flawed, in a group of politically pure midgets. Deprived of its intellectual justification and of those who had brought it into being, the great experiment that was to unite all people for all time beneath the banner of science was on the verge of disappearing in confusion.

For scientists in particular there must have been something particularly grotesque about the decision that winter to elevate reason to the status of a religion, at the very time when the Académie had been abolished, Lavoisier thrown in prison, Condorcet threatened with arrest, and science in general was disparaged. In May 1794 a tribunal would send Lavoisier to the guillotine with the contemptuous remark, 'The republic has no need of intellectuals,' prompting in turn Lagrange's memorably mordant eulogy, 'It took them only an instant to cut off that head, but it is unlikely that a hundred years will be enough to produce another one like it.'

That the idea of a decimalised system based on the meridian survived at all was mostly due to Maximilien Robespierre, the ideological perfectionist whose radical rhetoric and command of bureaucratic procedure earned him dominance over Jacobin party politics and control of the Committee of Public Safety. Although Public Safety gave him the political high ground, Robespierre's speeches showed that personally he placed a higher value on his membership of the Committee of Public Instruction, whose remit included not only education but anything that contributed to the general enlightenment of the citizen, including the new weights and measures. There could be nothing more purely rational than the metric system, based on science, universal in application, unencumbered by local variants, impeccably logical. 'Surely it demonstrates,' ran a Public Instruction report, 'that in this field as in many others, the French republic is superior to all other nations.' Adoption of the metric system was as patriotic a duty as wearing the three-coloured cockade.

In November 1793, Robespierre delivered a wide-ranging speech on foreign affairs, affirming France's solidarity with small, vulnerable republics, such as the Swiss cantons and the United States. There was little her conscript armies could do to aid the Americans, but French politicians were well aware that Congress was considering how to reform the country's weights and measures. Early in December, the Constituent Convention's President, abbé Grégoire, suggested that it would be a friendly gesture to share France's intellectual achievement of the metric system with the Americans, and the idea was enthusiastically taken up by the Public Instruction Committee. As its messenger, the Committee chose Joseph Dombey.

It was an unlucky choice.

Dombey was a fifty-two-year-old doctor, a botanist, and a man whose character was as playful and charming and sensitive as a kitten's. There is little doubt that it was largely the combination of his character and his scientific background that led to Dombey's selection. 'He was good-looking, with a charming smile,' wrote his friend André Thouin, one of France's leading naturalists. 'His eyes were large and dark, and the eyebrows thick and black. His skin was tanned, almost African in appearance. His character was gentle, open and sympathetic. In general he was high-spirited, although sometimes very low.' He had integrity, courage and a sense of adventure. He was the ideal choice in every way but one – his luck was phenomenally bad.

As a young man Dombey had been a rising star in the founding science of botany, and in 1778 the head of the royal gardens selected him to join a Spanish botanical expedition to Peru and Chile. His five years there earned him a nomination to one of the forty-two places in the Académie. A herbarium of plants he sent back to Paris contained 1500 new species and about sixty new genera. He brought back mineral specimens including thirty-eight pounds of platinum, discovered new species of hardwood in the headwaters of the Amazon, new shrubs in the Andes, and mercury and gold mines in Chile. He was attacked by Tupamaro Indians, collected Inca pottery, and described scientifically for the first time the magnificent 150-foot-tall Araucania pine.

Dombey should have been a hero, but pure misfortune destroyed all his achievements. One set of plants disappeared in a shipwreck, another was almost destroyed by Spanish officials, and worst of all, for a scientist, he was prevented from publishing his results. When he gave up botany and returned to doctoring, he chose to practise in Lyon, a city that Robespierre's Jacobins besieged and sacked in 1793. That summer the doctor's patients were hauled out of hospital for execution, the drains overflowed with blood running from the guillotines, and Dombey's mercurial spirits were so shattered that his friends thought he would go mad if he stayed in Revolutionary France. It was they who decided that he would make a suitable messenger to take the news of the metric system to the United States.

The idea quickly took root. Abbé Grégoire was told that Citizen Dombey, responsible for France's collection of South American plants, was anxious to create a similar collection from North America, and was willing to travel there at his own expense. It might also have been pointed out that he was a particularly suitable choice, since he had brought back the platinum from which the definitive *mètre* would have to be fashioned. Grégoire was convinced.

On 16 December, the Committee of Public Instruction issued orders for a replica *mètre* and *grave* to be made out of copper for Citizen Dombey, and on the same date Robespierre and six other members of the Committee of Public Safety ordered that a passport be issued to Citizen Dombey so that he could travel to the United States in order to tell Congress of the advantages of the metric system. He would also send back to France useful plants and seeds and obtain information about the United States in response to a series of questions formulated by the Committee of Public Instruction and the Natural History Society.

Early in January 1794, the short, once-jaunty figure of Joseph Dombey, now growing bald and sallow-skinned, set foot on the gangplank of the American brig the *Soon*. In his baggage he carried the means of measuring the world.

ELEVEN

Dombey's Luck

ON 17 JANUARY 1794 – the twenty-eighth day of Nivôse in Year II, by France's new partially decimalised calendar – Captain Nathaniel Williams Brown set sail from Le Havre with Joseph Dombey and his cargo, bound for Philadelphia. Swathed in linen and packed carefully in wooden boxes, the two copper objects looked like nothing useful, neither tool nor weapon nor kitchen utensil. One was a thin, flat bar, slightly over a yard long, or as French citizens would have said, about half a *toise.* The other resembled the little cylinder an apothecary or a spice merchant might put on a set of scales to measure out his goods, except that it weighed something over two pounds, a unit that no one on either side of the Atlantic had ever used. Locked into them was the most advanced science of the day.

Beside them, carefully folded into a leather pouch, was a one-page document containing a decree issued in Paris on 26 Frimaire (16 December 1793) and signed by, among others, Maximilien Robespierre, which explained the purpose of Dombey's voyage. It began: 'The Committee of Public Safety [considers] that it would be useful for the United States Congress to learn of the work of the Committee of Public Instruction on weights and measures . . .'.

All that Captain Brown from New York would have known about his passenger was that his voyage had official authorisation, but the New York newspapers followed every move of the French Revolution in detail, seeing in it a rerun of their own nation's fight for freedom. It would not have needed someone as astute as a Yankee skipper to guess that Citizen Dombey, with his impressive passport, was probably an emissary of some kind from France's new rulers, and that since he was bound for the capital of the United States his journey might have some political importance. Thus when Dombey came on board with his valuable sea-chests, Captain Brown would have been predisposed in his favour, and anxious to help him reach his destination as quickly as possible.

His passenger was a chunky, solid man, standing about five feet four inches, with a naturally lively face and, for all his hollow-cheeked fatigue, a charm that remained undiminished. Neither Brown nor Dombey had more than a smattering of the other's language, but events were to show that during the voyage that lay ahead the captain, like many before him, came to feel not just warmth, but deep affection for the Frenchman. Whether he also grew to sense, with a sailor's instinct, the bad luck he carried with him was too small a detail to catch history's attention.

There was little doubt that in Philadelphia too, Dombey would be warmly welcomed. Despite the execution of the King, and the Terror that Robespierre had launched against the enemies of the Revolution, France remained almost universally popular in the United States. She was still the ally whose ships, soldiers and money had helped win the new nation its independence. Since the summer of 1793 and the outbreak of war between Britain and France, Britain had imposed a trade embargo against France, which had the effect of reviving memories of the old Franco–American alliance. American grain ships heading for France were peremptorily halted on the high seas by the Royal Navy, and Congress resounded with furious speeches demanding retaliation. Even the Secretary of the Treasury, Alexander Hamilton, the firmest supporter of commercial ties with Britain, turned hostile, and gave the British Ambassador a tongue-lashing about the 'injuries which the commerce of this country had suffered'. However patronising Robespierre's view of the republic beyond the Atlantic might be, he could hardly have chosen a more opportune moment to offer it the metric system as evidence of France's friendly feelings.

Although Jefferson was no longer in office, his French sympathies and eagerness to meet foreign botanists would have led him to make Dombey doubly welcome in Monticello. Dombey would have been supplied with introductions to James Madison, James Monroe and Robert Livingston among the Republicans. He would have been bathed in that Congressional sympathy for the French which found expression in April 1794 in fulsome motions from both houses 'congratulating [France] upon the late brilliant successes of the arms of the Republic' against the British and their allies. President Washington, so concerned for the establishment of a uniform set of weights and measures to hold the young country together, would have approved the purpose of Dombey's visit, and promoted his cause among the Federalists. The sight of those two copper objects, so easily copied and sent out to every state in the union, together with the weighty scientific arguments supporting them, would have clarified the minds of senators and representatives alike. The vibrant, determined personality of Dombey would have created an immediate empathy. And today the United States would not be the last country in the world to resist the metric system. Would have, would have . . .

On the *Soon*, Dombey could expect to reach Philadelphia at the end of February 1794, on the eve of his fifty-third birthday. He had escaped the carnage of the Terror, he was serving the cause of science, and he had been given a second chance of making his name as a botanist. He had every reason to be happy, but his past had taught him the wisdom of an observation made by his hero, Linnaeus, the founder of modern botany. 'You see that when a man walks in the bright sunshine,' Linnaeus once said to an audience of Stockholm students, 'he has a double shadow following closely on his heels ... the darker shadow is Fortune, and the brighter one Virtue, these two accompany us everywhere across this world's arena, whether we walk or ride.' Virtue, he explained, was the urge to work hard and be a good scientist, Fortune was what determined whether anything came of all the effort. 'It is of no avail that one discharges one's duties well,' said Linnaeus, 'all depends on Time and Fortune.'

When they were no more than a fortnight out from Philadelphia, a storm caught the *Soon* and drove her south towards the Caribbean, forcing Captain Brown to put into the French colony of Guadeloupe for repairs. If that was bad luck, it was nothing compared to being caught up in the politics of the island, which was sharply divided between royalists and revolutionaries. Fired by Dombey's presence, the leader of the revolutionaries distributed a poster attacking 'that vile despot' General Victor Collot, the royalist Governor of Guadeloupe. In response, the Governor ordered Dombey's arrest as a trouble-maker, at which a crowd of infuriated citizens threatened to attack a royalist village in retaliation.

As the mob advanced towards the Salt River, a sea inlet separating them from the village, Dombey's bedrock qualities of courage and fairness emerge so powerfully that it is easy to imagine what a powerful advocate the Committee of Public Safety had chosen for their new system of measurement.

Standing alone on the bank, blocking their crossing, he tried to reason with the mob, and to convince them that bloodshed was not the answer. When those in front would not listen, he shouted to those behind that he was prepared to go voluntarily to see the Governor. Still the crowd would not turn back, but nor would Dombey budge, until the physical pressure of bodies forced him off the bank. He fell several feet into the Salt River, and the surge of waves through the narrow inlet quickly swept him away from the bank. A boat was launched to pick him up, and when he was eventually rescued he was unconscious.

What should have been a scientific mission was turning into operatic tragedy. For a few days more while he recovered from his near-drowning, Dombey continued to have the illusion of freedom, but in reality what would happen to him was inexorable. It was not in his character to break his promise to go voluntarily to Collot, and on 26 March he was taken round the coast to the island capital of Basse-Terre, where the Governor took him into custody until his future could be settled.

With the *Soon*'s repairs complete, Captain Brown might have sailed for the United States. The Caribbean was not a healthy place. A British fleet under Admiral Sir John Jervis had just sailed into the West Indies, and British naval captains had a habit of stopping American ships to search for seamen who had deserted. They would certainly remove a French national. There were pirates in these waters too and, more dangerous still, the sea-guerrillas, the privateers whom every maritime nation licensed to prey on the shipping of their enemies, and of neutrals suspected of helping them.

The sooner Brown was away, the better his chance of escaping detection by the British. And if he were stopped, Dombey's presence would be a liability, providing ample justification for a boarding party to regard his ship as hostile. It could only have been from friendship for Dombey that Brown heeded Governor Collot's summons and on 30 March sailed round the coast to Basse-Terre to pick up his passenger. Recognising the danger, Dombey disguised himself as a Spanish seaman – although it is hard to imagine what kind of pirate would mistake an elderly man with the pale skin and soft hands of a scientist for a sailor.

Brown was ready to sail the moment Dombey's chests were swung aboard, but he was forced to wait until a Swedish schooner under Collot's orders left the harbour first. Spectators on the steep slopes of the Soufrière volcano behind the harbour of Basse-Terre noticed that as the schooner reached open water, two privateers appeared over the horizon. Instead of turning away, the Swede altered course towards them and appeared to exchange some message before sheering off. Then the *Soon* emerged from the harbour, caught the wind in her sails and steered north for the United States. The privateers did not even wait for her to lose sight of land. While the hillside audience watched, they closed in, forcing her to heave-to, and as she lay rocking in the swell a prize crew was sent aboard.

Some friends of the revolutionaries on Guadeloupe claimed to have seen the royalist Governor Collot up there on the slopes as well. When the *Soon* was boarded, they alleged, he turned to one of his associates and remarked with satisfaction that Citizen Dombey was a plotter who had been taken care of. The British, he said, would certainly hang him.

That was not to be Dombey's fate. The privateers carried him and his belongings to the nearby British colony of Montserrat. He was a valuable prize, too valuable to hang. Among his papers were the diplomatic codes to be used by the French legation in the United States, for which the British would pay well, and the *Soon* together with its cargo could be sold as a prize. Captain Brown would be returned to an American port, and Dombey could wait in prison in Plymouth, Montserrat's capital, until France was prepared to negotiate for his release. For the privateers it was a good day's work. As for the copper bar and the weight, they would be auctioned off with the rest of the brig's cargo.

Dombey did not survive long in his sunless stone cell. In early April he died, and was buried on Montserrat. Today his grave, and much of Plymouth, is hidden under ash and rock thrown out when the volcano above the town began to erupt in the mid-1990s. Yet remarkably the two bronze measures which had cost him his life did reach the United States.

Sometime in July the cargo of the *Soon* was auctioned off, probably in New York or Boston, and the metre and the *grave*, together with the message from Committee of Public Safety, were bought by a French sympathiser who sent them on to the French Minister in Philadelphia, Joseph Fauchet. On 2 August, Fauchet presented them to the new Secretary of State, Edmund Randolph. Neither man was a scientist, and without Dombey to explain, they failed to appreciate the significance of the two standards. Indeed the prototypes of the metre and the kilogram were never shown to Congress at all. Somehow they survived two centuries of bureaucratic neglect and, by the sort of good fortune that escaped their bearer, today rest in a display case in the National Institute of Standards and Technology (NIST) in Washington. NIST is the government department which now spearheads official efforts to bring the metric system to the United States. In that sense, Dombey's copper bar and weight have still not arrived.

That winter, however, the Committee of Public Safety's message did reach the President. In January 1795 Washington communicated it to Congress with another appeal for action on the neglected question of weights and measures. Belatedly the House of Representatives took up the challenge, and appointed its own committee to consider both the President's appeal and Jefferson's scheme. In contrast to all previous reports, which had gone straight in to recommend a new system, this committee restricted itself to declaring some basic and by now familiar principles: that change was needed, and that a new system should be derived from the second's pendulum and the cube of rainwater. It did not commit itself to a decimal system, but simply asked that $1000 be allocated to conduct experiments to determine which system would be most convenient.

Unlike the Senate at that period, the House of Representatives allowed its debates to be published, and the opinions expressed in its discussion of the report give a good idea of those in the public at large. Some were for change and some against. Others abstained, saying that only scientists could understand the problem. When Jonathan Havens of New York earnestly tried to explain the science behind the second's pendulum, his fellow New Yorker William Cooper jumped to his feet and mockingly declared that it put him in mind of the lines from Oliver Goldsmith's 'The Deserted Village':

> While words of learned length and thundering sound
> Amaz'd the gazing rustics rang'd around;
> And still they gaz'd, and still the wonder grew,
> That one small head could carry all he knew.

But a sense of gravity was restored when the tycoon John Swanwick of Pennsylvania pulled himself up to his scant five feet, and said that he had known of too many 'disputes which have arisen for want of some certain standard to regulate weights and measures, and frequently the payment of a whole cargo [has been] disputed on account of a difference in the size of bushels'. Thomas Jefferson had always accepted that it was the business community who were most hostile to change, and this call for reform by a founder of the Bank of New York, and an unquestioned leader of the city's financial interests, represented a significant shift in attitudes. Swanwick was supported by Jonathan Dayton, who wielded a double influence as land speculator and Speaker of the House. The subject of weights and measures was very important, Dayton insisted, 'and to no country more than the United States, as every state has its own different weights and measures which causes the greatest uncertainty in all commercial transactions'. The commercial argument was one every Congressman understood. The jeers and dissent were silenced, and the House adopted the report unanimously.

Quite suddenly, it appeared that the French example had, however belatedly, given the political process the kick-start it required. A Bill was submitted and passed by the House. Under the title 'An Act directing certain experiments to be made to ascertain uniform standards of weights and measures for the United States', it was sent to the Senate. There it sailed through two readings and the committee stage, and on 31 May 1795, the last day of the session, the Senate informed the House that it would consider the Bill at the next session.

At which point it vanished. The next session came and went, and the Bill on weights and measures was never seen again.

The explanation lay with events taking place far away from Philadelphia, in the forests beyond the Ohio river. After two years of cautious fort construction far to the west where the Alleghenies start to flatten into the Indiana plains, the federal army under General Anthony Wayne had succeeded in encircling the forces of the Western Confederacy. In the summer of 1794 they advanced northward along the Maumee river towards the shores of Lake Erie, and in a clearing made by a tornado on the west bank of the river, they encountered the enemy. From defensive positions behind the tangle of storm-felled trees, Wayne's troops fired almost unscathed into the Confederacy's lines until return fire slackened, and then with a final bayonet charge won an emphatic victory.

The battle of Fallen Timbers brought the leaders of the Wyandot, the Delaware, the Miami and other Indian nations to make peace the following year at the Treaty of Greenville. There the Confederacy was forced to cede the remainder of Ohio and part of what is now Indiana to the United States.

Two other treaties were signed in that year of 1795 – the Jay Treaty with the British, who finally agreed to evacuate Detroit and other positions on the northern frontier, and the Treaty of San Lorenzo with the Spanish, which granted Americans freedom of navigation on the Mississippi and the use of New Orleans as a port. For the first time since independence the United States was free of external threat.

Once the menace of Indian attack, of British harassment,

of Spanish blockade, was removed, the log-jam that had held up settlement at the Ohio suddenly gave way. Behind the outriders of surveyors and squatters, an irresistible army of settlers pushed the Wyandots, the Miamis and their allies out of Indiana, out of Illinois, into one treaty after another, further and further west, until their identity almost disappeared. The window of opportunity for linking reform of the system of weights and measures to the sale of public domain land had closed.

On 18 May 1796, displaying an alacrity conspicuously absent where weights and measures were concerned, Congress passed 'An Act for the sale of land of the United States in the territory northwest of the River Ohio, and above the mouth of the Kentucky River'. The period of experimentation was over. The land would be surveyed by the government before sale, and there would be no metes and bounds, no squares alternating between townships and sections, no land company sales. The definitive pattern of six-mile-square townships, divided into one-mile-square sections, was established. Lines would run north–south, and east–west, and one section in each township would be reserved for education.

'The lines will be measured,' the Act carefully specified, 'with chains containing two perches of 16½ feet each, divided into 25 equal links, adjusted to a standard kept for that purpose.' The fact that a perch is also a pole and a rod may be confusing, but the authoritative gloss given by the historian of the survey, Albert White, makes the Act's import clear: 'This specifically calls for a Gunter's chain, and leaves no doubt that accurate measurements are to be made.' On a far wider scale of significance, this is also the first official United States unit of measurement. Congress might not be able to make up its mind about any other weight or measure, but it was determined that the dimensions of Gunter's chain should apply to the entire nation.

On the frontier and elsewhere, decimals were dead.

In June 1846, Charles Dickens travelled from London to Switzerland to begin work on a new novel which had a plot but no title. His journey took him through Macon, the town in eastern France where Joseph Dombey had been born. The family still lived and worked there, and the name, possibly glimpsed on their grocery store or in an advertisement for it, must have lodged in Dickens' memory, because a fortnight after arriving in Switzerland he wrote delightedly to a friend, 'BEGAN DOMBEY!' The character of Mr Dombey, frigid with ambition for himself and his family, could not have been further from that of the warm, mercurial Joseph, but so far as the wider world was concerned, the fictional Dombey was the only one who ever existed.

TWELVE

◆❋◆

The End of Rufus

THE *PITTSBURGH GAZETTE*, always a reliable guide to the economic health of the Ohio valley, reported in October 1795 that 'The emigration to this country this fall surpasses that of any other season – and we are informed that the banks of the Monongahela are lined with people intending for the settlement on the Ohio, and Kentucky. As an instance of the increasing prosperity of this part of the state, land that two or three years ago was sold for ten shillings [$2.50] per acre, will now bring upwards of three pounds [$15].'

To investors in the Ohio Company who had bought at about twelve cents an acre in 1785, the surge of settlers pushing up land values at last brought healthy profits. For Rufus Putnam in particular it represented a triumph of planning and persistence, and in November 1796 he was rewarded by being appointed Surveyor-General of the United States, with responsibility for surveying the rest of the Northwestern Territory.

In the aftermath of the Greenville Treaty, the great fort at Marietta was torn down and its red timbers used to construct more elegant, peaceful dwellings, the finest of them housing Rufus and Persis and their children. Looking round their home today, there is nothing incongruous between the belongings they had brought from Massachusetts – the fine bow-fronted walnut desk, the splendid four-poster bed, the deep-toned cello – and the handsome building with its dark, gleaming floors and its smooth, dazzlingly white plaster walls. Civilisation had leaped across the Ohio.

Rufus was now in his late fifties, and seemingly as energetic as ever – after the treaty he personally undertook the survey of the reserves which the Moravian missionaries had persuaded the government to set aside for them and their converts among the American Indians. As the plain, two-roomed land office built at the back of his house indicates, he remained brusque and straightforward in his attitude to the business of surveying. He contracted with Israel Ludlow to run the line marking the boundary specified by the Greenville Treaty between the Western Confederacy and the American settlers. It zigzagged south-westwards from the mouth of the Cuyahoga on Lake Erie to the mouth of the Kentucky river on the Ohio. Ludlow, as might have been expected, managed to bend the line by over five hundred yards. Reporting this to Putnam, he expressed the hope that the Indians would not notice, and since both agreed that it was a risk worth taking, the mistake remained uncorrected.

In surveying terminology, the crook in Ludlow's line was known as a jog, as though his elbow had been nudged while drawing it on the paper. The plats produced by the teams of surveyors working for Putnam were filled with jogs. He had divided the huge area north-west of the river Ohio – effectively central and western Ohio – into districts and assigned them to different survey teams, but the east–west parallels in one district rarely matched up with those of its neighbours. The north–south meridians added to Putnam's difficulties. Although the 1796 Act had specified that the meridians were to be run true north, the business of getting a fix on the Pole Star, and correcting for its deviation and for local magnetic variation, took too much time. None of Putnam's teams bothered with it, and he did not insist. Consequently the enormous Military Reserve, for example, which the United States had set aside for its veterans in Ohio was tilted four degrees off true north. Nor did anyone have time, at $3 a mile, to solve the problem of converging meridians, and sporadic corrections meant that sections that were supposed to contain 640 acres might vary in size from under six hundred acres to over seven hundred.

At the end of three years, nearly all the land inside the treaty line had been surveyed. Some of the jogs were very large indeed, but a pattern of (roughly) six-mile-square townships had been laid across the land, and the bristly old soldier Rufus Putnam had established a system of issuing contracts for specific surveys that all his successors were to follow. In Ohio, no one thought the worse of Rufus for his strange spelling, or for putting his son and a son-in-law on the federal payroll, or for his standards of surveying. He was regarded with affection, and respected as someone who got things done.

Other surveyors took the same pragmatic attitude to their job. In 1796 Moses Cleaveland, surveyor and major stockholder in the Connecticut Land Company, was running straight lines through the area which stretched north of Thomas Hutchins's Seven Ranges to Lake Erie. This was the 3.5-million-acre Western Reserve, which Connecticut had thoughtfully required the federal government to award it in return for giving up her charter claims in the west. The land company had bought it for $1.2 million, a sum that in good New England fashion was earmarked for education, and proved to be the foundation for Connecticut's superb public education system – but the company now needed a return on its money.

It was all good land, as the description of the Moravian missionary John Heckewelder made clear. 'Although the country in general containeth both arable Land & good Pasturage,' he wrote, 'yet there are particular Spots far preferable to others.' The best of all in his opinion was the place where the Cuyahoga river flowed into Lake Erie. It had a safe harbour, and the fishing there was good, Heckewelder reported, because it was 'a place to which the White Fish of the Lake resort in the Spring in order to Spawn'. The river itself 'has a clear & lively current but all Waters & Springs emptying in the same prove by their clearness & current that it must be a healthy Country in general'. Cleaveland was in search of a suitable location for the capital of the Western Reserve and, attracted by Heckewelder's magical description, took a surveying party and began to mark out a city among the trees.

For anyone unaware of Gunter's chain, the dimensions of what would eventually be Cleveland, Ohio, are hard to understand: the largest road, Superior Street, measured 2640 feet long and 132 feet broad, while minor streets were ninety-nine feet in width. But for chainmen working in dense woodland, it made sense to keep things simple, hence a main street forty chains long by two chains broad, and lesser roads one and a half chains across. The main square was ten acres in area, which happened to be ten chains by ten chains, and most of the city lots were two acres in area, or in surveyors' terms two chains broad by ten chains long.

Wherever speed was necessary, surveyors would 'gunterise' the measurements in this fashion – the 'National Road' constructed on Congress's orders across central Ohio and Indiana in 1796 to link settlers with their markets in the east was also ninety-nine feet broad – because it was quicker to keep things simple for the chainmen. For them as for Rufus, the priority was finishing the job rather than being fancy.

With the beginning of the new century, however, it became clear that in the public lands survey this rough and ready approach would no longer be tolerated. After the election of Jefferson as President in 1800 and his appointment of the ferociously clever, Swiss-born Albert Gallatin as Secretary of the Treasury, the messages from the nation's new capital, Washington, to Marietta became increasingly peremptory in tone. Rufus Putnam had never thought highly of Jefferson, whose leadership of the Republicans he regarded as disloyalty to George Washington. President or not, he would always be 'the Arch Enemy'. He was now about to be given a more personal reason for disliking Jefferson. Rufus was in his sixties and, in his attitude to surveying, quite evidently a holdover from the eighteenth century. In 1803 Gallatin sacked him – 'Because,' Rufus commented bitterly, 'I did not die nor resigne' – and without acknowledging his achievements brusquely instructed him to turn over his papers to the new Surveyor-General, Jared Mansfield, a former professor of mathematics at West Point military academy.

Rufus responded to this shabby treatment with dignity, handing over the records but writing to Gallatin in a tone which offered a hint of the roaring lion and tusked boars on his coat of arms: 'Perhaps you may imagine this conduct looks like passive obedience and non-resistance, or that I am courting favor. Mistake me not. I have done no more than what I conceive to be the duty of every public officer . . . I am too independent to be influenced by the prejudices of the times.'

It was Rufus's last public growl as Surveyor-General, and it is to be hoped that he was soothed by the sound of Persis's cello, and the sight of their fine furniture, and the solidity of their fortress-timbered house, and the satisfaction of having his grandchildren playing round him in the garden above the Muskingum. Rufus and Persis passed the remainder of their days in Marietta, and by the time Rufus died in 1824, four years after his wife, the land survey had been pushed westward across Indiana and Illinois and stretched from Lake Michigan south to the Gulf of Mexico.

The changes introduced during those years by Rufus's two successors, Jared Mansfield and Edward Tiffin, showed why he had had to be replaced. It was not just the prejudice of the times that led to his sacking – it was a fear that the entire survey might fall apart.

The trouble stemmed from the uncontrollable surge in land

speculation. More than in any other economy of the time, American land was the prime producer of wealth, partly from its crops and livestock, but mostly from the increase in its value. The mathematics was spelled out crudely by a spokesman for the North American Land Company, which had invested in unoccupied land west of the Appalachians. The population had doubled in the last twenty-five years, the spokesman said in 1790, and would again in the next twenty-five. 'Supposing half of each state to be unoccupied,' he explained, 'it follows that in twenty-five years there would be an increase of inhabitants sufficient to settle this vacant half. The average price of the lands in the settled half of the United States cannot be at less than eight dollars an acre – sixteen times the price at which the lands of the company are [bought] for its shareholders.'

The figures were tempting, and everyone with spare cash invested in land. The President did, Supreme Court justices did, Congressmen, Senators and Governors did. 'All I am now worth was gained by speculations in land,' the new Secretary of State, Timothy Pickering, told his sister in 1796. 'In 1785 I purchased about twelve thousand acres in Pennsylvania which cost me about one shilling [about fifteen cents] in lawful money an acre . . . The lowest value of the worst tract is now not below two dollars an acre.'

One of the heaviest plungers was Judge James Wilson of the Supreme Court, who at the urging of his friend Silas Dean, an associate of Robert Morris, took every chance to acquire land however dubious the circumstances. 'If we review the rise and progress of private fortunes in America,' Dean advised him, 'we shall find that a very small proportion of them has arisen or been acquired by commerce, compared with those made by prudent purchases and management of lands.'

The trick was not to wait for demand to push up the price, but to buy, as Pickering did, with devalued military land warrants. These gave the holder the right to a set number of acres, ranging from a hundred for a private to as many as

1500 for a general, at a nominal price of $1 an acre, but in the 1780s they could be bought for twenty cents on the dollar or less. The profit margin could be increased further by persuading the seller to give a discount for quantity. Up to 1800 the largest sellers of public land were the states rather than the federal government. Like the United States itself, they too faced debts arising from the war with Britain, and so once the surveyors had run lines establishing their borders, each individual state started to sell off territory to fill up its depleted treasury. Between 1783 and 1800 the states sold around fifty million acres, and the same names appear repeatedly as buyers: the Morrises, Robert and Gouverneur, Alexander Macomb, William Duer, and William Bingham of Pennsylvania.

There was nothing illegal about buying up military warrants, or even about persuading the New York legislature to sell over three million acres for as little as eight cents an acre, as Macomb did, despite the smell of bribery hanging in the air. Free enterprise was born out of land dealing, and long before the first business corporation existed, land companies issued shares and created many of the financial and legal structures that the nineteenth-century stock-dealing, capitalist economy used to finance the railroads and industrialisation of the United States. As early as 1765, Patrick Henry created the first pure trust, the North American Land Company, a legal ploy he devised to protect the assets of none other than Robert Morris. Financial institutions in Boston, Philadelphia and New York issued prospectuses and evolved increasingly sophisticated financial structures to tempt people to invest in land. Many of the largest transactions were handled by banks in Britain and the Netherlands which had access to even larger sums. So close was the involvement of London financiers in the land dealing that netted William Bingham some four million acres from Maine to North Carolina that Bingham's daughter Anne Louise eventually married Alexander Baring, the head of Barings Bank. Their expertise in the American property market would soon enable Barings to take on the financing needed for the most gigantic land deal of all – the Louisiana Purchase.

One other participant in the Purchase also gained his first experience of the United States from speculators. When Talleyrand appeared in New York in 1794, fleeing from the Terror in France, he ought as the apostle of French metric reform to have visited Thomas Jefferson, but he had a stronger tie with Gouverneur Morris. In 1785, when Gouverneur was negotiating tobacco deals in France for his business colleague and namesake Robert, he and Talleyrand had shared the same mistress, Adèle de Flahaut. At the time, fashionable circles speculated gleefully on the sexual orientation that led her to choose Talleyrand, who had a club foot, and Gouverneur, who stumped around on a wooden leg; but none of the three cared. 'My friend's countenance,' Gouverneur wrote of Adèle, 'glows with Satisfaction in looking at the Bishop and myself as we sit together agreeing in Sentiment & supporting the Opinions of each other.'

Naturally, in the United States Talleyrand attached himself to Gouverneur, and thus entered Robert Morris's circle. He soon found himself despatched to Maine to spy out new lands for purchase there, and as he surveyed the grandeur of the landscape his thoughts became unmistakably American. 'There were forests as old as the world itself,' he wrote, 'green and luxuriant grass decking the banks of rivers: large natural meadows, strange and delicate flowers . . . in the face of these immense solitudes we gave vent to our imagination. Our minds built cities, villages and hamlets.' Failing to make a living from real estate, Talleyrand eventually returned to France, but in 1803 his experience would produce for the United States an unexpected windfall.

Where speculation grew dangerous was the point at which northern financial sophistication mixed with southern metes and bounds. A metes and bounds survey did not just produce shapes that only the best surveyors could measure, it created a maze of bureaucratic form-filling that invited fraud and wholesale corruption. In theory the purchaser handed over to the state treasurer the money or military warrant for the amount of land he wanted. The receipt was taken to the land registry, which issued another warrant, which had to be taken to the county surveyor, who laid off the land that was wanted and gave the purchaser a certificate describing the property. The certificate was returned to the land registry so that the property could be patented, and only then was a document finally issued proving that the purchaser was indeed the owner of the land.

In practice the procedure was complicated further by the inaccurate maps drawn up by poorly trained surveyors, and the mistakes made freely or for bribes by inadequately paid registrars and land officers, so that legal claims were forgotten or pre-dated. Sometimes disputes were settled by force on the land itself, sometimes in court, and occasionally by free admission, as in 1816 when Kentucky's auditor contritely revealed that hundreds of legally purchased farms had never been registered, and that as a result the state had sold the owners' land all over again. Absentee buyers found it advisable to employ local land-jobbers to protect their investments by running fresh surveys, registering their claims and suing in the county courts. The qualities they looked for were those of Uria Brown, a well-known Virginia jobber who in 1816 promised his clients that he would 'Appear solid & firm, & persist in Establishing the rights of Lands as the Tuffest skin shall hold out the Longest . . . & surveys on surveys is there, [k]nee Deep and deeper.'

In effect the metes and bounds system was skewed in favour of those with deep enough pockets to hire lawyers and land-jobbers, and to keep sweet an army of state officials. These were the very people who could afford to invest in depreciated warrants or paper money. Robert Morris had shown how it was done in 1783 with his speculative purchase of more than a million acres in Virginia, paid for in military warrants. When more land was released in 1792 in exchange for heavily discounted Virginia Treasury notes, the speculators descended like vultures and bought up the notes, backed by credit from brokerage houses in Boston, Philadelphia and New York. A total of 2.5 million acres ended up in the hands of just fourteen individuals, most of whom were absentees. George Washington himself pointed to the consequence in western Virginia, where 'the greater part if not all the good Lands, on the main river, are in the hands of persons who do not incline to reside thereon themselves, and possibly hold them too high for others'.

Speculators also had the inside track when the political decision was taken to sell land. North Carolina's sale of four million acres in 1783 enabled the well-connected William Blount to acquire over a million acres of them, mostly in Tennessee, and when he was appointed Governor, he in turn helped his political friends to more land. Blount at least was a resident, but in an exhaustively researched thesis, *Speculators and Settler Capitalists* (1995), Wilma A. Dunaway suggests that at this period absentee landowners held title to three-quarters of the southern Appalachian lands. Much of what remained was owned by a tiny aristocratic minority. Just 4 per cent of the land, the most mountainous and least productive part of it, remained to be divided up among 80 per cent of the population, the majority of whom were left as landless tenants, sharecroppers or slaves.

So much empty land owned by absentees was a recipe for resentment. Landless residents burned the blazed trees that marked boundaries, or moved monuments, and occasionally entirely new plats were forged. To substantiate their claims they relied on local knowledge, and on home-grown specialists known as 'red brush surveyors' who knew every wrinkle of the land and could recognise boundaries by memory. But unlike the northern states, the south rarely allowed any rights of pre-emption to squatters or 'trespassers having no color of title' until the 1820s.

The often violent struggles that broke out between the absentees and the jobbers trying to hold onto their vacant land, and the squatters and red brush surveyors trying to move in, had an unintended consequence. Doubts over who actually owned properties slowed the market in land. In 1816 Uria Brown warned that speculators 'would purchase no Lands in [west] Virginia at any price: for the Titles of Land there was worse than the Titles in Kentucky'. And to show how bad that was, he prophesied, 'the titles in Kentucky w[ill] be Disputed for a Century to Come yet, when it's an old Settled Country'. In contrast to the north, where families would settle, improve, sell up and move on in a restless fury that astonished European visitors, Southerners once they had bought their land tended to stick there, at least partly because legal doubts restricted the land market to local buyers and sellers.

The consequences of those distant metes and bounds surveys and their associated complications continue to mark the rural economies of southern Appalachia. Modern studies of eighty counties in the area show that, with uncanny persistence, over half of their twenty million acres are owned by 1 per cent of the tax-paying population, and that almost three-quarters of the owners are absentee. Power and money still rest with the mineral, timber and agricultural concerns whose enormous holdings occupy the most productive ground. The contrast with the subsistence farmers barely surviving on smallholdings up in the high woodland is as stark as it was two centuries ago.

A more benign legacy exists in the relationship of Southerners to their land, closer it usually seems than in the north. It ought, after all, to be easier to identify with a property whose boundaries are the streams and ridges chosen by metes and bounds surveyors, than with the inhuman rectangles of the public survey. Sometimes tortured, sometimes sublime, especially where the property has stayed in the family for generations, the intensity of such feelings infuses the work of Southern writers like William Faulkner and Eudora Welty. 'Don't you see?' exclaims Ike McCaslin in a fine gothic frenzy in Faulkner's *Go Down Moses*, 'This whole land, the whole South, is cursed, and all of us who derive from it, whom it ever suckled, white and black both, lie under the curse?' The impact of metes and bounds surveys and flawed land titles on Southern literature would be worth studying.

It was in Georgia that land sales crossed the blurred line between legal complication and outright fraud. The corruption rewarded a state governor, federal and state legislators, and justices up to the United States Supreme Court, and created in Georgia a culture that one modern analyst has compared to modern Russia, where powerful cartels aim not only to break the law, but to use it to legitimise their criminal activities.

Alone among the original states, Georgia had refused to cede to the United States her charter claims to land in the west, which included much of present-day Alabama and Mississippi. The Appalachian mountain range that separated other states from their western territory only skirts the north-west corner of Georgia, and there was no physical barrier to prevent settlers moving west until they reached the Mississippi. Consequently the state, the youngest and poorest of the thirteen, was eager to sell the land before it was taken up by squatters. Ever since South Carolina's plantation owners had moved across the border in colonial days and helped break down James Oglethorpe's surveyed plan of settlement, Georgia's land sales had been anarchic. Now they became a corrupted feeding ground for speculators.

In 1789 three different companies were formed by syndicates from Tennessee, South Carolina and Virginia – this last headed by Patrick Henry – and persuaded the state to let them buy twenty million acres centred around the Yazoo river valley for $207,000. Most of the territory belonged to the Cherokee, Choctaw and Chickasaw, and according to federal law could only be bought by the United States government. The audacious deal was shot down, not for this reason, but because the companies attempted to pay in depreciated certificates, a piece of sharp practice that left Patrick Henry's shining image tarnished in the state. Nevertheless the seed was sown.

In 1795 Spain relinquished her claims to Georgia's western territory, and four new companies were promptly formed to acquire land now undeniably American. Within months both houses of the Georgia legislature had passed legislation selling them forty million acres, much of modern Alabama and Mississippi, for just $500,000, although only half that sum was actually paid. An official inquiry by James Madison later discovered that all but one of the legislators had an interest in the sale, either in shares or bribes.

Public outrage voted in an anti-sales party which overturned the agreement thirteen months after it was made, but the damage had been done. Yazoo land had already been sold on for over $1.5 million, and among the secondary purchasers were Judge James Wilson of the Supreme Court, who bought 750,000 acres, William Blount, Governor of the Tennessee Territory, members of the United States Congress, and, almost inevitably, Robert Morris. Claiming that Georgia's agreement to sell the Yazoo lands was a valid contract, they took their case to law, and eventually to the Supreme Court. In 1809 Chief Justice John Marshall, a friend of Robert Morris and an advocate of company rights, held that the sale, despite its fraudulent circumstances, was indeed a contract. The case, Fletcher vs. Peck, not only won the Yazooists compensation, but underpinned the entire land market that was to fuel the United States's westward expansion. The nation's 'manifest destiny to overspread the continent' that John L. O'Sullivan discerned in 1845 could hardly have been fulfilled had destiny not been able to cite Fletcher vs. Peck.

In Georgia, it was merely its size that made the Yazoo fraud extraordinary. The state was so plagued by chicanery in selling land – the Surveyor-General's office was discovered to have a box full of plats of imaginary thousand-acre properties which only required compass bearings and dates to be filled in to become apparently legal documents – that in 1805 it resorted to the disposal of public land by lottery; only to discover that the lotteries themselves were rigged in favour of the speculators. Of the North American Land Company's listed assets of six million acres, four million were in Georgia, and there must have been widespread satisfaction when many of them proved to consist of fake plats – legal arguments about what Robert Morris had actually bought were still being heard as late as 1882.

In 1803 what concerned Albert Gallatin and Thomas Jefferson was the way that fraud and confusion were spilling out of South Carolina and Georgia into the federal public lands. The 1790 law creating the Southwestern Territory, which consisted of Tennessee and, after 1802, of Alabama and Mississippi, required the land to be surveyed, sold and governed on the same basis as the Northwestern Territory. The practice was different. In the north, something like a grid was being imposed on public land; in the south, metes and bounds became the norm, and with them came corruption, litigation and a land market that rewarded the speculator and the banker. For a President whose political philosophy centred on widespread ownership of land, it was intolerable.

The surveyor of the south, Isaac Briggs, a talented engineer who would soon be involved in the successful survey of the Erie Canal, evidently froze at the tangle of problems before him. The heat, the fever, the belts of heavily wooded, swampy terrain, all made surveying so difficult that even at a special rate of $4 a day he could not find enough men prepared to take on the work. The surveyors he did hire attempted to lay out townships on the northern model, but found they had to deal not only with squatters but with a slew of existing properties bought through the Yazoo land companies and other speculative investors, or by private arrangement with Spain, or, especially in regions like Natchez, Mississippi, where many Americans had acquired Cherokee land, by deals with the native inhabitants. Each had to be separately surveyed and entered on the plat – 'a severe duty which the surveyors complain of', explained Briggs's successor, Thomas Freeman. There were so many that whatever the law might demand, their plats had the disordered appearance of a metes and bounds survey.

Determined to prevent Southern land patterns taking over, Gallatin wrote impatiently to Briggs in 1805: 'It is true that you will not be able to complete your work in that scientifick manner which was desireable, [but] it is of primary importance that the land should be surveyed and divided, as well as it can be done.' It was advice that would have drawn a grim smile from Rufus Putnam.

The date of Gallatin's letter was significant. A few months earlier, the last detail had been completed on the Louisiana Purchase, a deal which added 900,000 square miles of new territory to the 400,000 already in the public domain. It was no longer a matter of having Alabama and Mississippi properly surveyed. What would happen to all the land that the United States now owned west of the Mississippi river?

The Louisiana Purchase took place because the priorities of both sides made a perfect fit – Napoleon's need for money to finance the next phase of his European war was matched by the United States's requirement to control the port of New Orleans at the mouth of the Mississippi – but it was significant that on each side the negotiators were utterly familiar with the workings of the land market. Talleyrand had worked for Robert Morris and had dreamed the speculators' dreams, and

First Furrow by Olaf Carl Seltzer (1877–1975). The dispossessed view the possessor. Unseen are the surveyor's squares that made the transfer possible.

An early-nineteenth-century view of Washington, D.C., the city that Pierre L'Enfant designed with avenues radiating outwards from Capitol Hill and the White House in uncommercial stars and circles.

Joseph Dombey (1742–1794), sculpted by Jean-Antoine Houdon.

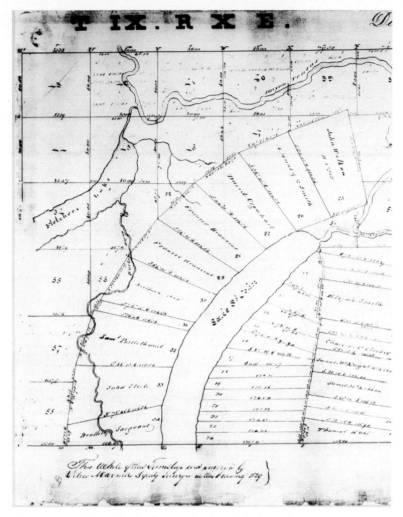

The squares of the public land survey (eighty chains by eighty) come up against the long lots (roughly forty *arpents* by five) of French settlers on the Red River in Louisiana.

An 1850 map of San Francisco, a neat grid of squares laid across precipitous slopes and rugged hills. The scale is given in the original Mexican *varas* as well as English feet.

Opposite Western San Francisco, where the grid that Jasper O'Farrell drew spectacularly over the distant hills of the city spreads neatly towards the Pacific.

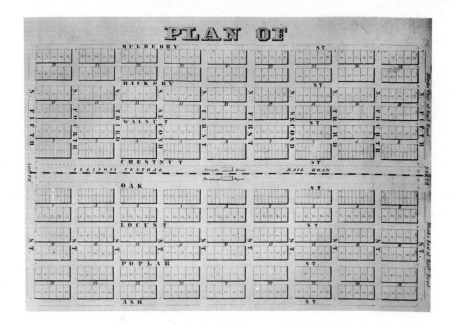

PLAN OF

Above The Illinois Central railroad's standard town plan, creating hundreds of identical towns across Illinois. Further west, thousands of cloned cities grew from similar plans drawn up by other railroads.

Right The business end of the survey. A Kansas Land Office in 1874 with a county map on the back wall squared into sections, and a potential purchaser being shown a pre-surveyed plot.

Joseph Smith's idea of heaven. The City of Zion – also known as Salt Lake City – pictured in 1870. Each square contains ten acres or ten chains by ten, dimensions taken from the Book of Numbers.

The Oklahoma Land Rush. The start of a race to stake claims to land that had belonged to American Indians and was about to become American property.

A survey team over thirty strong in 1871 near what would become the United States's first national park at Yellowstone in the Rockies.

The 1871 team under John W. Powell (standing on the central boat) that surveyed the Colorado river valley. A well-protected theodolite stands on the shore.

Robert Livingston, the American Minister in Paris, owned the massive Clermont estate on the upper Hudson and knew every large speculator in New York. The arrival in Paris in April 1803 of James Monroe as Jefferson's personal representative simply added a Piedmont Virginian steeped in a tradition of land acquisition.

Consequently, when Talleyrand suddenly offered American negotiators not just New Orleans but all of France's territory beyond the Mississippi, the bizarre nature of what was happening – the sale of an empire – was not an issue; it was simply a land deal. The sum asked, $15 million, exceeded by $5 million the maximum that Monroe and Livingston were authorised to spend, but that was less important than the price per acre. Although neither side knew the exact size of French Louisiana – 'I can give you no guidance, but you've made a good deal for yourselves,' Talleyrand said airily – the two Americans could work out that with public domain land selling at $2 an acre there had to be enough profit in the deal for them to close without referring the offer back to Jefferson. (It turned out that they were paying less than five cents an acre.) Neatly tying up the business, the finance for the transaction was provided by Alexander Baring of Barings Bank in London, who as William Bingham's son-in-law was familiar with wheeling and dealing in American real estate.

Once it belonged to the United States; the Louisiana Purchase needed to be divided up and sold on, but every problem that Isaac Briggs had encountered on the east bank of the Mississippi was to be found on the west. The United States undertook to honour property registered with the French and Spanish authorities, but the promise constantly threatened to break the land survey. In Orleans Territory, around New Orleans, the long lots of French farmers, measuring five *arpents* by forty (about 320 yards by 2600) and running back from rivers and creeks were so numerous they forced surveyors to abandon six-mile squares altogether wherever there were waterways.

Silas Bent, the Surveyor-General of Louisiana Territory, comprising modern-day Arkansas and Missouri, found more sinister problems. Wherever there was good land, squatters claimed to have French or Spanish land grants, and they not only blazed trees but also forged the records. The registry office files 'have undergone a revolution', Bent reported to Gallatin in 1806. 'There has been Leaves cut out of the Books and others pasted in with Large Plats of Surveys on them . . . the dates have been evidently altered in a large proportion of the certificates. Plats have been altered from smaller to Larger. Names erased and others incerted and striking difference in collour of the ink etc . . .'.

There was a large and growing threat that all the corrupt practices and inefficient surveys that dogged land distribution in the southern Appalachians would become endemic west of the Mississippi. The entire administration of the public lands survey had to be overhauled. Surveys had to be closely controlled, their accuracy improved, and a system of checking imposed. Above all, the standards expected of the surveyors had to become professional. That was why rough and ready Rufus Putnam had had to go, and why Jared Mansfield had been appointed Surveyor-General of the United States.

THIRTEEN

The Immaculate Grid

THE REGULARITY THAT Jared Mansfield imposed on both himself and the United States was ruthless. From the moment he left Yale, his life as teacher, soldier and instructor at West Point military academy was one of such blameless rectitude, it is almost a relief to discover that he was in fact expelled from Yale for 'discreditable escapades'. Yet even here his crime – cheating in an exam – might be seen as an attempt to achieve perfection. Certainly the undeviating straightness of the rest of his career must have served as some form of compensation for the one stain on his character.

Mansfield was a mathematician of the highest class, and his *Essays, Mathematical and Physical*, published in 1802, was the first contribution to that field by a native-born American. Astronomy and the calculations needed to establish earth positions from celestial observation received particular emphasis in his writing. It was noticeable too that, following his appointment as Surveyor-General, Mansfield placed an order with leading instrument-makers in London for 'A three-foot Reflecting Telescope, mounted in the best manner, with Wollaston's *Catalogue of the Stars*, Maskelyne's *Observations and Tables*, A thirty inch Portable Transit Instrument, answering also the purpose of an Equal Altitude Instrument and Therdolete [*sic*], [and] An Astronomical Pendulum Clock'. This was not the action of someone who would tolerate curving straight lines and squares that looked like diamonds.

The best way to appreciate what Mansfield did is to drive west from Jonathan Dayton's speculative city on Interstate 70, and just across the Indiana state border to swing north on State Route 277. Beyond the wide verges of the Interstate, a beautiful rolling landscape can be glimpsed between predatory trucks and glaring billboards, but once on the narrow strip of 277, you become part of that country. The road acts as boundary to fields of wheat and soya, lawns with picket fences back onto it, and it is shaded by stands of oak and maple and walnut. The geological *tsunami* of the Alleghenies has virtually subsided here, leaving only ripples of rock beneath the surface. Sometimes blacktop, sometimes grey gravel, 277 runs straight as a surveyor's ruler over the ripples, giving a ride as exhilarating as a powerboat at sea. In the troughs, all you can see are the nearest red-painted barns and green John Deere harvesters, but from the peaks the distant spires of churches and silver grain-towers appear above the surging land.

But Route 277 has another claim to distinction. Exactly one mile to the east, Indiana's border with Ohio runs parallel to it, a north–south line which Jared Mansfield designated as the First Principal Meridian. West of that line he was to establish a survey whose squares were so immaculate that their pattern would be compared to graph paper, checkerboards and plaid. In other words, Ohio with its different and indifferent surveys was the proving-ground for the system, but 277 marks the first line inside Mansfield's monumental, continent-wide grid-iron.

Its regularity grew from the initial point, which was formed at the meeting of the principal meridian – a carefully surveyed north–south line – and an east–west base-line crossing it exactly at right angles. The squares were numbered outwards from that zero point: running east–west they were called ranges, and north–south they were townships; thus the first square west of a meridian and north of a base-line would be titled Range 1 West, Township 1 North. Mansfield himself personally surveyed the Second Principal Meridian, whose initial point can still be found a few miles south of Paoli, Indiana, and others were run as new areas of land were prised from the Native Americans and put on the market. Instead of Rufus Putnam's independent survey districts, the different areas could be connected by extending a base-line or a meridian. The most spectacular example was the Fifth Principal Meridian, whose initial point was in Arkansas, near the present town of Blakton, twenty-six miles west of the Mississippi, but which was extended so far north that it eventually controlled other land surveys in Arkansas, Missouri, Iowa, South Dakota, North Dakota and most of Minnesota, ending with Township 164 North, on the Canadian border.

Mansfield's method also provided a solution to the problem of converging meridians. The curvature of the earth brings lines of longitude gradually together as they run towards the pole, so that in most of the United States the northern end of a township is thirty to forty feet narrower than the southern. In Alaska the flattening of the earth means that the gap closes by over a hundred feet (some of Israel Ludlow's township meridians in Ohio converged by more than three hundred feet, but that was not the earth's fault). As townships were stacked on top of each other, they grew narrower, and after four or five, the most northerly might be over sixty yards narrower than it was supposed to be. At that point Mansfield's successor, Edward Tiffin, decreed that a fresh start should be made, with new meridians exactly six miles apart marked off on the base-line. The jog created by these 'correction lines', where the old north–south line abruptly stopped and a new one began fifty or sixty yards further west, became a feature of the grid, and because back roads tend to follow surveyors' lines, they present an interesting driving hazard today. After miles of straight gravel or blacktop, the sudden appearance of a correction line catches most drivers by surprise, and frantic tyre marks show where vehicles have been thrown into hasty ninety-degree turns, followed by a second skid after a short stretch running west or east when the road heads north again onto the new meridian.

Despite these improvements, Mansfield's surveyors were not immune from error. Massive jogs occurred in eastern Illinois, where the parallels running out from the Second Principal Meridian spectacularly failed to meet up with those of the Third Principal Meridian, and a 'shatter zone' of steeply angled lines had to be introduced to join one lot to the other. But such mistakes were inevitable given the pressure the surveyors were under to make land available for sale as quickly as possible.

Rapidly though the squares were laid out, it was not quick enough to satisfy the press of settlers. The westward drive was fuelled partly by the rising population – the 1800 census showed that in ten years the number of Americans had grown by 35 per cent to 5,306,000 – but especially by the mouth-watering prospect of the land beyond the Alleghenies, beginning roughly at Chillicothe, whose ancient mounds proved that it had been productive ground centuries before the first European landed. For the land-hungry, timber-yearning, field-dreaming squatters in Kentucky and Virginia, those gentle slopes, rich meadows and rolling oakwoods beyond the hills were the stuff of dreams. Once clear of the mess of claims and counter-claims around the speculative city of Dayton, this country stretched out empty and inviting as far as the sterile, unforested prairies.

As early as 1798 the federal government in Philadelphia had sent Rufus Putnam an anxious request for information about a group of three hundred Kentucky families who had settled in unsurveyed country just beyond the treaty line. When he went to investigate, he found that many had come intending to buy the lands 'as they should be offered for sale by the United States', but that rising prices had pushed them on. Even in Dayton the price of $2 town lots with uncertain title were, according to the speculator John Symmes in 1796, 'selling in Cincinnati at ten dollars per lot'. Cheaper land could only be found further west. The race that developed between the surveyors and squatters marked the entire history of the land survey, and it was rare for a surveying team to measure productive country which had no settlers at all.

In their desire for land, pioneers like John Pulliam, who had already squatted on farms in Virginia and Kentucky, leap-frogged far ahead of the tide. In the breakout year of 1796 he crossed the Ohio, and with his family moved on through what would become Indiana and Illinois as far as the Mississippi. Much of this was compacted prairie soil, where tall bluestem grass and wild rye grew taller than a horse's head, and their roots knotted into an unyielding mass that could not be broken up until John Deere's heavy, self-scouring steel ploughs were introduced in the 1840s.

The Pulliam family were Scotch-Irish, and as stubborn and restless as their eighteenth-century forebears who had swarmed unchecked through the Virginia piedmont in the 1730s to the fury of William Byrd. With their few goods piled in a cart, accompanied by long-legged hogs and some scrawny cattle, they settled for a year or two on land beside any creek where there was enough water to produce a stand of timber. Even shallow-rooted elms and sycamores broke up the ground well enough for it to be ploughed, and gave timber for making cabins. Along with maples, whose sweet sap in springtime could be boiled into sugar, there were oaks and beeches that produced acorns and mast for the hogs, and hickories like the pecan whose nuts could be pounded to flour or eaten whole, and fruit trees and wild vines carrying ripe cherries and sweet grapes. This produce ensured that the earliest settlers were timber-dwellers rather than grassland farmers.

The story of the Pulliam family was collected by nineteenth-century antiquarians because John's descendants were the first to move into Sangamon County in Illinois, but as John Mack Faragher commented in *Sugar Creek*, his 1986 history of that community, it is likely that John Pulliam 'spent his whole life farming without ever owning land'. The reason was simple. The 1796 Act specified that the land was to be sold by the section, that is 640 acres, and at a minimum price of $2 per acre, which, with costs for the first year's subsistence and improvements like fencing and building, left little change from $2000. To someone like John Pulliam, whose last days were passed as a ferryman on the Kaskaskia river, this was an impossible sum.

Successive pieces of legislation rapidly reduced the smallest amount of land that could be bought. In 1800 it was a half section (320 acres) and the down-payment was only a quarter of the total price, the rest being paid over the next three years. Congress authorised the sale of land by the quarter section (160 acres) in 1804, and in 1820 by the half-quarter section (eighty acres), with the price reduced to only $1.25 an acre. The changes had their effect, and Robert Pulliam, John's son, took the opportunity to buy the land that he farmed around Sugar Creek in Sangamon County. A shooting accident cost Robert his right foot, and his temper grew evil, but that did not prevent him acquiring more acres than his father ever squatted on.

The beauty of the land survey as refined by Jared Mansfield was that it made buying simple, whether by squatter, settler or speculator. The system gave every parcel of virgin ground a unique identity, beginning with the township. Its name might be Township 2 North, Range 4 West, Second Principal Meridian. Within the township, the thirty-six sections were numbered in an idiosyncratic fashion established by the 1796 Act, beginning with Section 1 in the north-east corner, and continuing first westward then eastward, back and forth in 'boustrophedonic' fashion, that is like an ox pulling a plough, until Section 36 was reached in the south-east corner.

Each square-mile section consequently had its own identity, and as the size of the minimum parcel of land shrank, the names simply grew more specific. In 1832, the smallest area for which a would-be farmer could bid at a government land auction was reduced to the quarter-quarter section, or forty acres, a parcel that has entered American rural mythology. The surveyors hated it, complaining bitterly of the paperwork involved in 'the new and minute subdivisions of fractional sections', but in the jargon of modern geographers it would become 'the modular unit of settlement'. It was the minimum area that was needed to support the average family. Railroads sold land by the forty-acre lot. After the Civil War, freed slaves were reckoned to be self-sufficient with 'forty acres and a mule', and in the twentieth century real-estate developers preferred to deal in forty-acre parcels. Even rotary sprinklers were designed to irrigate forty acres of grass. And long before the United States Postal Service ever dreamed of zip codes, every one of these quarter-quarter sections had its own address, as in ¼ South-West, ¼ Section North-West, Section 8, Township 22 North, Range 4 West, Fifth Principal Meridian.

What the system depended upon was a clear title to the property. This was indeed the very purpose of the grid. As a result, there was always a strand of opinion which regarded squatters as a menace. In 1807 a draconian Act authorised their punishment with fines and imprisonment, but it was rarely enforced, and after years of turning a blind eye to the practice, the Pre-emption Act of 1841 gave squatters the legal right to do what up to then had been a custom backed by local squatter power: that of buying the land they had improved by ploughing or building a cabin, at the going rate of $1.25 per acre. But long before then, most squatters had recognised the power of a surveyor's plat.

The desire to possess land drew people westward, but it was the survey that made possession legal. Three years after Robert Pulliam began farming in Sugar Creek, the Deputy Surveyor for that district, Angus Langham, arrived with his chainmen and axemen, who were also referred to as moundmen in the woodless prairies because they constructed mounds rather than blazing trees at section corners. These were a different generation from the surveyors who had worked for Rufus Putnam, and their growing professionalism could be seen in the instruction booklet issued to each of them.

First the township or exterior lines were to be established. From the initial point, the surveyors were instructed to move westward along the base-line parallel across the entire district, marking it with posts at every half-mile and mile for the quarter-section and section lines, as well as with six-mile markers for the township corners. From each of the six-mile markers a meridian was then run north, marked in the same way. Finally the team returned to the initial point and worked their way north along the principal meridian. At each six-mile point, they stopped and ran an east–west parallel to cut across the north–south lines. Branded onto every corner mark was the number of the section, the township, and range.

Along the way surveyors had to make a map on a scale of four inches to the mile, and keep notes in their field-book, recording the distance covered and the principal natural features found. And in case they did not know, the manual offered this example of how to record a meridian running north from a base-line:

> *North* Along the east boundary of Section 36, Township 21 north of the baseline, Range 6 east of the 4th principal meridian.
> Chains
> *14.70* A brook, 25 links wide, with a rapid current, runs south westerly about 10 chains, then turns to the N.W.
> *27.60* Left the creek bottom and entered hills.
> *29.40* A white oak, 15 inches diameter.

33.70 A hickory, 24 inches diameter.

40.00 set a quarter Section corner post on the top of a ridge, bearing north easterly and south westerly; from which post a white oak bears S[outh] 28 [degrees] W[est] 197 links, and a poplar bears N[orth] 56 [degrees] W[est] 14 links distant. The soil is good and fit for cultivation; timber walnut, cherry and white oak; undergrowth pawpaw and spice.

49.07 A white oak

64.08 A walnut

80.00 Set a post, corner to Sections 25 and 36, Township 21 north, Range 6 East of the 4th principal meridian; from which a hickory bears South 57 degrees West 127 links; and a white oak bears North 23 degrees West, 72 links distant. Land too hilly for cultivation, although the soil is rich; timber, hickory, white oak and walnut; undergrowth pawpaw and spice.

That represented one side of one square-mile section, and the same procedure had to be repeated for all thirty-six square miles of the township. Then, beginning at the south-east corner of each township, they worked west and north filling in the section lines. Shrewd speculators learned to avoid land in the north-west corner, because any errors in measurement showed up there, and that was where most arguments over boundaries occurred.

When the surveyor's plat was delivered to the district land office, it would show that every square of prairie or corner of forest had been given an identity – where Robert Pulliam's farm stood it was Section 21, Township 14 North, Range 5 West, Third Principal Meridian, together with the letters 'AP', meaning applied for, which showed that the land had been pre-empted. Eventually the claim had to be paid for, and once that was done it was entered on the definitive survey map and patented, at which point the prairie or forest became private property, whose ownership would be protected by the full force of the law. This was something the wildest squatter eventually learned. As a means of defence, a surveyor's plat carried more firepower than a Kentucky long rifle packed full of lead shot and black powder.

Once that was understood, the influence of the grid stretched out to places long before the surveyors arrived. Anyone who intended to settle and buy land would try to mark out his claim in rectangles so that it aligned with the grid. In 1832 one nameless pioneer who had moved far ahead of the survey into the Iowa prairies explained how he did it: 'The absence of section lines rendered it necessary to take the sun at noon [to find north–south] and at evening [to find east–west] as a guide by which to run these claim lines. So many steps each way [eight hundred double-paces by 1600] counted three hundred and twenty acres, more or less the legal area of a claim. It may readily be supposed that these lines were far from correct, but they answered all the necessary claim purposes for it was understood among the settlers that when land came to be surveyed and entered, all inequalities would be put right [by adding or subtracting land].'

What this makes clear is how easy it was to measure out land using the grid and the old four-based organic measures. Acres had evolved from the basic productive needs of human society, and in the primitive conditions of the frontier they again came into their own. When Congress made it possible to sell public land by the quarter-quarter section, the squatter had only to step out 250 double paces, or 440 yards, towards the sun at noon, stick in a marker, then march 250 double paces towards the point where the sun set – 'only alternate steps are counted', W.F. Horton reminded readers of his *Landbuyer's, Settler's and Explorer's Guide*, published in 1902 – and put in another marker. That was it: 440 yards by 440 made a forty-acre lot. Even Jefferson would have approved – no system of measurement could have been more transparent, more democratic, more suited to 'the calculation of everyone who possesses the first elements of arithmetic'. It was so straightforward that the citizen squatter could operate it as easily as the government surveyor.

As territory south of Tennessee was squeezed out of the Cherokee, Choctaw and Chickasaw, Mansfield's system was applied there as well. Two principal meridians were established, the Washington Meridian in south-east Mississippi, and the St Stephen's Meridian in southern Alabama, both taking as their base-line the thirty-first parallel, the boundary between the United States's and Spain's territory. Yet even under Mansfield's rigid rules, Northern regularity still battled to overcome Southern customs.

In the first two decades of the nineteenth century, cotton prices doubled, and profits fuelled a boom in production which began around Natchez then spread all along the lowlands east of the Mississippi. Plantation owners in Virginia and the Carolinas began to sell up their exhausted land and move to Alabama and Mississippi accompanied by their slaves, livestock, furniture and machinery. Since this was an undertaking on the scale of relocating a factory today, and often financed by investors in New York and Philadelphia, care had to be taken to acquire the best site.

The soil most suited to cotton-growing was the rich valley-land, most of which was Cherokee-owned, although some was squatted, or the claims poorly documented. To acquire legal title to it amid the confusion of existing private claims required inside knowledge, which wealthy purchasers could buy from the public land office in Huntsville, Alabama. In 1818, a year after General John Coffee was appointed Surveyor-General of Alabama, the *Huntsville Republican* informed its readers that the land office would 'give any information to people wishing to purchase an advantage', in return for 'a liberal per centum' of the price. Whether the client wished to locate a suitable tract or to engross several different parcels, if necessary removing squatters, Coffee's employees, the newspaper alleged, were prepared to 'do business on commission, and receive in pay either a part of the land purchased; or money'.

The society that evolved from this pattern of land distribution seemed far less exotic to foreign visitors than the one in the North. The South's large estates and social privilege produced a hierarchy that nineteenth-century Europeans found familiar, except for its reliance on slavery. Many were shocked by the discovery that slavery was still legal in the United States. What they failed to appreciate was that in the new nation the concept of property had evolved beyond their experience into something absolute, that overrode even the founding principle that all men were created equal with the inalienable right of liberty. Slaves were property – 'For actual property has been lawfully vested in that form,' Jefferson bleakly wrote, 'and who can lawfully take it from the possessors?' – and against that their humanity counted for nothing.

Thanks to Rufus Putnam's ban on slavery, the new states in the North had escaped that evil, and in their energetic, egalitarian materialism, European visitors found the utterly foreign experience most of them were looking for. It was based on owning the land that Mansfield had measured out for the settlers. 'The possession of land is the aim of all action, generally speaking and the cure for all social evils among men in the United States,' wrote the visiting English writer Harriet Martineau in *Society in America* in 1837. 'If a man is disappointed in politics or love, he goes and buys land. If he disgraces himself, he betakes himself to a lot in the west. If the demand for any article of manufacture slackens, the operatives drop into the unsettled lands. If a citizen's neighbors rise above him in the towns, he betakes himself where he can be monarch of all he surveys.'

It was so obvious, so widespread that few Americans recognised that it was remarkable, but European visitors immediately found in the pattern of land ownership something quite distinct from the hierarchical, landlord-dominated, agricultural societies they were accustomed to. One of the earliest accounts, from John Melish, a political radical who first travelled through the west in 1806, shows clearly that he saw land ownership as the key to American independence of spirit. 'Every industrious citizen of the United States has the power to become a freeholder, on paying the small sum of eighty dollars, being the first installment on the purchase of a quarter of a section of land,' he wrote in *Travels in the United States*; 'and though he should not have a shilling in the world, he can easily clear as much from the land, as will pay the remaining installments before they come due.'

This was optimistic. Land office records were filled with entries of farmers who had failed to keep up their payments. But Melish had no intention of letting detail obscure the broad picture he wanted to paint, because it was unlike anything known to his readership in Britain. 'The land being purely his own, there is no setting limits to his prosperity. No proud tyrant can lord it over him – he has no rent to pay – no game laws – nor timber laws nor fishing laws to dread. He has no taxes to pay except his equal share for the support of the civil government of the country, which is but a trifle ... Such are the blessings enjoyed by the American farmer ... May the Almighty Father of the human race pour down his choicest blessing on the heads of those who planned, and carried into effect such a benevolent system.'

Not surprisingly, Jefferson read Melish's book 'with extreme satisfaction'. It takes a small effort to appreciate how odd the American system would have appeared to most of Melish's readership, in whose experience the privilege of owning land freehold was largely confined to the aristocrat, the squire and the gentleman. Here it was available to anyone prepared to work hard enough to earn the $160, and from 1820 only $100, price of eighty acres. 'I *own* here a far better estate than I *rented* in England,' wrote Morris Birkbeck in the 1820s, 'and am already more attached to the soil. Here, every citizen, whether by birthright or adoption is part of the government, identified with it, not virtually but in fact . . . I love this government.' Birkbeck wrote to pull in immigrants to the Illinois colony he had founded, but the emotional connection he made between land ownership and democracy was unmistakably genuine. This was the bright side of the coin called property.

When Captain Frederick Marryat, a bluff naval officer and prolific author of best-selling adventure stories, visited Galena, Wisconsin in 1837, he could hardly believe that such privileges of ownership were supported by the law even against the government, and even when they were claimed by a squatter. As an example, he wrote of the federal depository which had been built at Galena to hold the lead that miners in the territory paid as tax. 'As soon as the government had finished it,' he recorded, 'a man stepped forward and proved his right of pre-emption on the land upon which the building was erected, and it was decided against the government, although the land was actually government land.' Marryat's tone was amused rather than admiring, much as when he wrote of the reaction of backwoods squatters to attempts to evict them, 'the consequences were very commonly that the new proprietor was found some fine morning with a rifle-bullet through his head'. It was all part of the wild and woolly west, he implied, and not something that any well-run society would want to copy.

The French politician and writer Alexis de Tocqueville was so struck by the system of land ownership in the United States that he made the mistake of asserting in *Democracy in America* (1835–40) that: 'In America there are, properly speaking, no farming tenants; every man owns the ground he tills.' In fact much land was leased or rented, but on a small scale compared to the pattern of tenant-farmers in Britain, and landlord and peasant in France and Germany.

Unlike John Melish, the wonderfully opinionated observer Fanny Trollope, mother of Anthony, was a Tory to the tip of her parasol, but from the opposite end of the political spectrum her conclusions exactly matched his. They were prompted by a visit to a pioneer farm in the forest near Cincinnati in 1828. The family lived in a log cabin they had built beside a river, with a peach and apple orchard at the back, and cows, horses and pigs nearby. They grew potatoes on land they themselves had cleared, wove and knitted their own cotton and woollen clothes, made soap and candles, went to market to sell butter and chickens, and buy tea, coffee and whiskey, and were, she admitted 'indeed independent'. But they had achieved this rural idyll, she decided, at the cost of cutting themselves off from social life as she understood it: 'they pay neither taxes nor tythes, are never expected to pull off a hat or to make a courtesy, and will live and die without hearing or uttering the . . . words, "God save the king"'.

The social consequences of such independence were quite clear to her, as she wrote in *The Domestic Manners of the Americans* (1832): 'Any man's son may become the equal of any other man's son, and the consciousness of this is certainly a spur to exertion; on the other hand, it is also a spur to that coarse familiarity, untempered by any shadow of respect, which is assumed by the grossest and lowest in their intercourse with the highest and most refined.'

For good or ill, a new kind of society was evolving from the way in which the public land was being measured out. It could be seen in Congress, where a perceptible change in policies and attitudes occurred as representatives of the new western states, like the future President William Henry Harrison, a delegate of the Ohio Territory, took their places.

In his influential 1893 essay 'The Significance of the Frontier in American History', Frederick Jackson Turner identified this change as the point when the United States took on the individualistic, egalitarian qualities he associated with the frontier spirit: 'The frontier promoted the formation of a composite nationality for the American people . . . In the crucible of the frontier the immigrants were Americanised, liberated, and fused into a mixed race.' As a result, frontier people had a sense of themselves as American and an instinct for democracy that they had to teach the Atlantic states, which were still ridden with sectional self-interest. 'The rise of democracy as an effective force in the nation,' Turner asserted, 'came in with Western preponderance under [Andrew] Jackson and William Henry Harrison, and it meant the triumph of the frontier.'

Attractive though it is, Turner's argument has been repeatedly demolished by historians and geographers pointing out its inherent inaccuracies. Westward expansion was always piecemeal and irregular rather than consisting of a coherent moving frontier. Far from being individualistic, it was usually communal, and often built around urban rather than rural settlements. Above all it was fuelled more by speculation than the desire for liberty. Yet for all the academic scorn, the thesis refuses to die, simply because a distinctive, utterly American spirit did indeed arise from the expansion into the west. To most foreign observers, the origin of that spirit was obvious. It had nothing to do with the frontier family's encounter with the wilderness, and everything to do with their acquisition of landed property.

Ownership encouraged more than a sense of independence. Divided up into squares, the land that reminded the earliest colonists of God, and provoked a sensual hunger in later settlers, could also be treated by speculators simply as a commodity defined by numbers. A uniform, invariable shape that took no account of springs or hills or swamps was an obstacle to efficient agriculture, but to a financier tracking the rise and fall in land values, it was a great convenience.

The grid, designed by Thomas Jefferson to create republican farmers, also turned out to be ideal for buying, trading and speculating. The consequence was what D.W. Meinig, doyen of American geographers, termed 'the most basic feature of the settlement process: that it tended to be suffused in speculation'. The paradox was that most of the speculators were not big-time financiers – though they were there in plenty – but small-time republican farmers. Speculation, or as the nineteenth century called it, capitalism, and democracy went together.

It took time for the movement to acquire momentum. In the first three decades of the century, sales rose from about 300,000 acres a year to around a million; but then the process suddenly accelerated. Between 1830 and 1837, more than fifty-seven million acres were sold. The General Land Office, which had been established in 1812 to supervise the survey, was almost overwhelmed, but the sheer size of the market was itself proof of the effectiveness of Jared Mansfield's squares. A sudden collapse in prices in 1837 and again twenty years later put momentary brakes on sales, and the discovery of gold in California in 1848 created a west coast acquisition of land before the mid-west was half-settled. Nevertheless, by the end of the nineteenth century more than a quarter of a billion acres of public domain had been converted into private property.

Caught by the romance of the frontier, it is easy to miss what underpinned its spirit of restlessness, individualism and enterprise. The adventure of taming the wilderness was certainly there, but what drew people from eastern states and from around the world was the desire for this soil magically transformed from wilderness to property by the act of measurement and mapping. The fact that the grid could fractally subdivide a continent to minute, graph-paper squares might appear to be simply a triumph of the mathematician's art, but the ease with which it made land available to anyone who went west in search of it had an almost incalculable influence on the development of the American economy and the American psyche.

FOURTEEN

---✦---

The Shape of Cities

THERE WAS A SIMPLE TRUTH contained in the land survey – that a shape which could be neatly reduced to smaller proportions was easy to sell. That had been apparent to John Jacob Astor as early as 1800 when he bought his first building on the Lower East Side of Manhattan. There was something disgusting in Astor's inability to disguise his greed. He shovelled food into his greasy mouth, scooping up peas and ice cream with his knife, wiping his hands on the table-cloth – and he gobbled up land in Manhattan with the same unseemly haste, spending over $200,000 in one year alone to swell a hoard that eventually numbered over three hundred plots. His American Fur Company made money by trading beaver-pelts in the far west, but nothing like the fortune he earned simply by waiting for his property in and around the growing city of New York to rise in value. Land bought in 1800 for $50 an acre had a price-tag of $1500 in 1820, and when he died in 1848 Astor was worth $25 million, two and a half times as much as the next-wealthiest American.

Underpinning the rapid expansion of both Manhattan and Astor's wealth was the plan drawn up by the City Commissioners in 1811 for New York's future growth. Most of the city was still concentrated into the southern tip of the island, below what is now Canal Street, although it was spreading northwards, fuelled by prosperity from trade through the port.

As the three City Commissioners explained in their report, they faced a crucial decision: 'whether we should confine ourselves to rectilinear and rectangular streets, or whether we should adopt some of those supposed improvements, by circles, ovals, and stars which certainly embellish a plan, whatever may be their effects as to convenience and utility'.

The example of ovals and stars the Commissioners were referring to was Washington, superbly laid out in 1792 by Pierre L'Enfant with the help of Andrew Ellicott and the African-American surveyor Benjamin Banneker; but in terms of development Washington, with its avenues radiating out from the Capitol and the White House, was a disaster. Despite the lure of being the nation's capital, there was no market in town-lots there. It was Washington more than anything that finally broke the old speculator, Robert Morris. Having acquired over seven thousand lots at $66.50 each, Morris counted himself lucky to sell five hundred of them, and ended up serving a six-year stretch in the debtors' prison in Prune Street, Philadelphia – which he airily described as 'the hotel with grated doors'. When the duc de la Rochefoucauld passed through some years after the city's foundation, he remarked that Washington was no more than some buildings scattered among the woods, and that 'most of them were built for speculation and remained empty'.

Since one of New York's Commissioners was Robert Morris's namesake and sidekick, Gouverneur Morris, it was certain that they would avoid Washington's example. 'In considering that subject [the city's shape],' they wrote, 'we could not but bear in mind that a city is to be composed principally of the habitations of men, and strait-sided and right angled houses are the most cheap to build and the most convenient to live in. The effect of these plain and simple reflections was decisive.' Their choice was a grid-iron of a dozen north–south avenues, each a hundred feet wide, crossed at right angles by 155 east–west streets. The distance between the avenues varied – the result of the existing city layout – but between the streets it was decided by the surveyor's standby, Gunter's chain. Every new Manhattan block was to measure exactly three chains or 198 feet deep. It created a thin block of between three and four acres, compared to the more usual five acres in cities elsewhere. Confined to so small an area, Manhattan developers would soon have no choice but to build upwards.

The Commissioners' solution resembled the land survey in ignoring natural features of the terrain, like hills and swamps, and in maximising the opportunity for speculation. Once the plan was adopted, it became easy to predict where the city would spread. As Astor explained when a customer asked why he was selling a Wall Street location for $8000 rather than holding on for a few years to get $12,000, 'See what I intend doing with these $8000. I shall buy eighty lots above Canal Street, and by the time your one lot is worth $12,000, my eighty lots will be worth $80,000.' To criticism that the Commissioners' plan was unimaginative, the city's surveyor, John Randel, would always point out that it was ideal for the 'buying, selling and improving of real estate'.

A process of simplification, that decimalisers would have appreciated, led the 198-foot depth of a block to be referred to as two hundred feet, although as late as 1879 Frederick Olmsted, pioneer of urban parks and open places, was complaining bitterly that if a building 'needs a space of ground more than sixty-six yards in extent from north to south, the system forbids that it shall be built in New York'. These regulations gave rise to the standard Manhattan lot, which was generally described as a hundred feet deep (but was in fact one foot shorter), backing onto another of the same depth, and twenty-five feet broad. The dimensions enabled nineteenth-century developers to cram up to sixty lots onto each floor, with only narrow, dark stairways and tiny shared sinks and toilets to separate them. By the end of the century, a typical lot in a five-storey tenement on the Lower East Side would be subdivided into four apartments, each paying around $10 a month rent, and housing ten or more people. Epidemics of typhus and typhoid fever regularly ripped through what was at that date the most densely inhabited spot on earth, but the inhabitants' welfare attracted less attention than the fact that Manhattan's grid made it possible for one of its meagre three-acre blocks to generate an income of $12,000 a month in 1900, a time when $350 would keep a family for a year.

Manhattan's plain, profitable shape was to influence the design of American cities all the way to the Pacific, in contrast to Washington, whose intricate, unsaleable pattern was rarely copied – the magnificent heart of Indianapolis being the prime exception. To twentieth-century town planners like Lewis Mumford, the grid-iron was a disaster, mixing up residential and industrial areas, creating traffic problems and hideous slums. In John Reps' superb panorama of city design, *The Making of Urban America*, published in 1965, the Manhattan plan is depicted as little better than the plague: 'The fact that it was this gridiron New York that served as a model for later cities was a disaster whose consequences have barely been mitigated by more modern city planners ... [It] stamped an identical brand of uniformity and mediocrity on American cities from coast to coast.' Since those words were written, a revolution in construction and design has earned Manhattan its present reputation as one of the architectural wonders of the world, but its ground-plan still causes snarl-ups, aids urban sprawl, and has helped create slums in cities across the nation.

In 1830 James Thompson, a surveyor and engineer, was commissioned to lay out a town in Illinois, in the square mile of Section 9, Township 39, Range 14, Second Principal Meridian, so that lots could be sold to finance the Illinois canal. He took the Manhattan grid-iron a step further in regularity by dividing the section into squares measuring twenty chains by twenty chains – the homesteader's forty – and then subdividing each square into four blocks, whose dimensions of ten chains by ten included both buildings and streets. As had happened in Manhattan and across the land survey, speculators were attracted by the ease of multiplying or splitting up square and oblong properties.

Six years later, Harriet Martineau saw the results at first hand when she visited the mosquito-ridden swamp that Thompson named Chicago. 'The streets were crowded with land speculators hurrying from one sale to another,' she wrote. 'A negro dressed up in scarlet, bearing a scarlet flag and riding a white horse, announced the time of sale. At every street corner where he stopped, the crowd gathered round him; and it seemed as if some prevalent mania infected the whole people.' For the next seventy years, Chicago's grid spread outwards in the undeviating pattern that Gunter's chain made so simple.

It was difficult for any urban plan to escape the influence, either direct or indirect, of the twenty-two-yard length that was dividing up the country. Even in Thomas Holme's carefully conceived 1682 plan of Philadelphia, ten chains by ten is the space allotted to the five great open squares located at the city's centre and four corners. It is noticeable that the street widths, carefully drawn in Holme's office, are 150 feet; but when Philadelphia's grid-iron pattern was adapted for use in other eighteenth-century cities, surveyors armed with chains found these decimal numbers hard to measure out on the ground. The solution was to gunterise them to ninety-nine feet or 150 links, and 49½ feet or seventy-five links.

This was what happened when Joseph Smith, founder of the Church of Jesus Christ of the Latter Day Saints, designed a settlement in Jackson County, Missouri, in 1833. The square shape was based on the instructions for city-building in Chapter 35 of the Bible's Book of Numbers, but the cubits in the original specifications were gunterised to rods and perches.

'We send by this mail a draft of the City of Zion,' Smith told the Church. 'The plot contains one mile square; all the squares in the plot contain ten acres each, being forty rods [ten chains] square . . . Each lot is four perches [one chain] in front, and twenty [five chains] back, making one-half acre in each lot.' To maintain the town's regularity, all the streets were to be two chains broad. After Smith's murder by a mob in 1844, following his arrest on charges of treason and conspiracy, these dimensions were followed for each of the cities built by the Mormons in their search for a safe haven, the last and most spectacular of them being Salt Lake City, which Orson Pratt laid out on the directions of Smith's successor Brigham Young beneath the Wahsatch mountains in Utah. Intriguingly, excavations for new sidewalks undertaken in Salt Lake City in 2001 revealed that the corner-posts were set four inches further out than they should have been, suggesting that Pratt had not followed the practice of public lands surveyors and calibrated his chain precisely before starting work.

There was nothing inevitable about these urban shapes. In New England the colonists had developed the village pattern, in which the original purchasers settled round a green at the centre of the community, with fields behind their houses, and grazing land still further out. In Montreal, Mobile and New Orleans, the French planners produced designs centred on a hollow square from which narrow streets ran out, lined by houses with gardens and courtyards behind. In the province of Quebec, they experimented with fortified villages at the centre of a star of wedged-shaped fields. The impact of the 1526 Law of the Indies, which did envisage a grid for settlements in Spanish America, was limited in its effect to the streets round the central square, and beyond that the typical *pueblo* usually developed in haphazard fashion, more in response to the availability of water than to planners' dictates.

Once the United States land survey established its pattern, however, it was hard to resist its rectangularity. The chain's length was so well-fitted to the section's 640 acres that few city planners could resist incorporating its dimensions into their designs, especially when speed and quick sales were required. Even the radiating design of Indianapolis had to fit inside a section's one-mile-square box, and the sad story of Circleville, Ohio, which was planned as a series of rings but was speedily converted to squares by the residents, proved how hard it was to break the pattern.

It was the railroads that made the most extreme use of the grid and the chain. As an incentive to build, the federal government awarded railroad companies blocks of land, usually a township but sometimes more, alternating on either side of the track. Most of it was sold off to farmers, but the real money lay in towns. Companies like the Northern Pacific and the Burlington designed a standard town that could be laid out and sold off wherever they decided to site a passenger and freight depot. The basic model consisted of three 160-acre sections on each side of the track, each section being split four ways into those forty-acre – twenty by twenty chains – lots that a surveyor could measure with his eyes closed. All the east–west streets were named after trees like Hickory, Walnut and Chestnut, and the north–south streets either numbered or given the letters of the alphabet.

'An office boy can figure out the number of square feet involved in a street opening,' Lewis Mumford exclaimed, 'and a lawyer's clerk can write a description of the necessary deed of sale merely by copying a standard document. With a T-square and a triangle, finally, the municipal engineer without the slightest training as either an architect or a sociologist could "plan" a metropolis.'

This of course was the virtue of the scheme in the railroad companies' eyes. Seven dollars was as much as the Burlington was prepared to pay to have a town planned. Sometimes the companies built only on one side of the tracks, in which case they simply cut the plan in half, but the depot always remained the focal point. Out west a depot was required every twelve or fifteen miles so that farms would not be more than half a day's journey away, and hundreds of identical towns were laid out with a railroad down the centre, two parallel streets on one side named Oak and Chestnut, and on the other side two more named Walnut and Hickory, crossed by ten others named First to Tenth. There were so many that it was hard to come up with names for them all.

'I shall have two or three more towns to name very soon,' Charlie Perkins, land agent to the Burlington railroad, wrote in the 1870s as the track advanced through Iowa. 'They should be short and easily pronounced. Frederic I think is a very good name. It is now literally a cornfield, so I cannot have it surveyed, but yesterday a man came to arrange to put a hotel there. This is a great country for hotels.'

The $7 plans were hyped in beguiling fashion. 'Would you make money easy?' asked George F. Train, a founder of the Union Pacific railroad. 'Find then the site of a city and buy the farm it is to be built on! How many regret the non-purchase of that lot in New York; that block in Buffalo; that acre in Chicago, that quarter-section in Omaha . . .'. This was the more honest approach, though Train himself died a pauper. The less honest and more common method was to pretend that the towns were already built and thriving. '*Towns* they are on paper,' confessed the amiable Charlie Perkins, 'meadows or timberland with here and there a house, in reality.'

It was not until the beginning of the twentieth century that the planners succeeded in fighting free of the sort of drab materialism the square encouraged. The turning point was the 1893 Chicago World Fair, where examples of spectacular and innovative architecture triggered such enthusiasm that Daniel Burnham, the founder of American town planning, was able to persuade the city authorities in Washington, Cleveland and Chicago to redesign public buildings in his trademark style of gleaming white stucco, decorated with domes and pillars and set off by parks and esplanades. When San Francisco was all but destroyed by earthquake followed by fire in April 1906, Burnham produced a new city plan with more parks, curving roads and diagonal thoroughfares to take the place of the plain grid hurriedly laid down in the days of the gold rush. As cities grew, suburban planners followed the advice given by Frederick Olmsted that they should adopt 'in the design of your roads, gracefully curved lines, generous spaces, and the absence of sharp corners, the idea being to suggest and imply leisure, contemplativeness and happy tranquillity'.

Yet even as this idea took hold, and developers began to create the new concept of suburban living which deliberately shunned the sharp right-angles of farms and city blocks, an old reality still lurked below the surface. The original real estate was still bought by the square forty-acre lot, house-plots were sold in ten-, five- or most commonly 2.5-acre parcels, and streets tended to measure sixty-six feet, or two chains, in width. As one Illinois developer observed in 1966, 'Underneath all these contemporary trappings, our basic thinking is still geared to a gridiron block system.'

That system's resilience was apparent too when San Francisco arose from the ashes of earthquake and fire. Instead of following Burnham's graceful curves and diagonals, the city kept to the old gold-rush grid that climbed straight up Nob Hill and hurtled straight down to the Bay. Even John Reps was prepared to admit that for once squares and straight lines could be exhilarating. 'San Francisco is a glorious exception to the otherwise gloomy record of the grid,' he wrote. 'On no site less superb could this have happened, nor is this, visually the greatest of America's cities, likely to be duplicated.'

Jasper O'Farrell, who laid out San Francisco's original grid in 1845, would have been gratified, but he would have noted too that the new plan designated the dimensions of a city block in feet and inches. The unit he had used to measure it out was the Mexican *vara*. Since then, a minor miracle had occurred: the American Customary System of Weights and Measures had become the law of the land.

FIFTEEN

Hassler's Passion

I N DECEMBER 1806, Thomas Jefferson's former ally in the reform of weights and measures, Professor Robert Patterson of the University of Pennsylvania, sent a letter to the President recommending the employment of a Swiss immigrant called Ferdinand Rudolph Hassler. 'In his education he paid particular attention to the study of astronomy and surveying,' Patterson wrote. 'He is a man of a sound, hardy constitution, about thirty-five years of age, and of the most amiable, conciliating manners.'

Even this glowing recommendation did not convey Hassler's unique quality. He was a geodesist, an earth-measurer. As a student at Bern University he had met Johan George Tralles, a gifted scientist and father-figure, who introduced him to the exquisite pleasure of land-measurement coupled with the harsh discipline of pinpoint accuracy, and the masochistic science captured Hassler for life. In 1791 he and Tralles undertook the first geodetic survey of the canton of Bern in his native country. The report of the Economic Society of Bern that led to the project emphasised the economic value of knowing the precise heights of hills and valleys, and the exact location of natural and man-made features, when planning major engineering works like roads and canals. But the difference between a geodetic survey and any other kind was not just a matter of economic advantage. A geodetic survey was so precise that it increased the supply of the world's knowledge about the shape of the earth and, in the words of the report, would 'secure the thanks of the world of learning and be a lasting honour'. The inextricable alliance of science and industry that would mark the nineteenth century was on its way.

Two years later Hassler and Tralles travelled to Paris, where they met Delambre, Lavoisier and other leading figures of the metric commission, and acquired copies of the metre and the kilogram. Both men became enthusiasts for what Hassler termed 'the best and most consistent system of weights and measures hitherto devised', and in 1799 Tralles would take a leading role in the international commission that verified the scientific basis of the metric system.

On his return to Switzerland Hassler fell in love with Marianne Gaillard, whose surname, 'gaiety', conveyed her light-hearted nature. She had been brought up to sing and play the piano, and to occupy the role of wife to a comfortable Swiss public servant. Her husband's personality, however, nursed a contradiction exhibited by so many land measurers that it is almost a defining characteristic – a passion both for exact definition and for untamed wilderness. When in 1803 the invading French took over the triangulation of Switzerland on which Hassler had been employed, the buried side of his nature came to the surface.

In 1804 he decided to emigrate, and start a colony in the south-east of the United States. Marianne's reaction is not recorded, nor her feelings on their arrival in Philadelphia when they discovered that the agent to whom they had entrusted their money had gambled it all on a fraudulent land deal and lost everything. But she may have found some compensation in the fact that poverty forced Ferdinand to stay in the city rather than going off to be a farmer in the country. Almost at once he became involved with the American Philosophical Society, of which both Patterson and Jefferson were members.

Among the objects packed into the ninety-six trunks and boxes Hassler brought with him were his kilogram and metre. To keep the family fed he was forced to sell these precious items and most of his instruments to John Vaughan, a member of the Philosophical Society; but his expertise in French decimal measurement was noted, and some metric enthusiasts asked him to survey a long enough stretch of the Hudson river valley to provide an American basis for the metre. The project came to nothing, but it persuaded several other members of the Philosophical Society to write to Jefferson about the remarkable Swiss scientist in their midst.

These letters initiated what turned out to be the only successful attempt to decimalise the United States's measures for which Jefferson could claim responsibility. Like earlier efforts, it was connected to a major programme of land measurement. In 1807 Jefferson persuaded Congress to authorise funding for the United States Coast Survey. As its name implies, the survey was intended to provide accurate maps of the eastern seaboard, which now stretched from Maine to New Orleans. They were needed for defence and for the huge maritime traffic which carried America's goods between the Mississippi valley and the Atlantic coast, and onward to the outside world – but these were merely the practical reasons. Together with Albert Gallatin, his Treasury Secretary, who had a grounding in mathematics and a commitment to science as firm as his own, Jefferson intended to give the scheme a wider context.

The geodetic survey of Switzerland was only part of an international trend to precision mapping. France led the way, and by the end of the century had been measured from top to bottom three times. When Delambre presented Napoleon with his acerbic memoir of the final, meridian survey, *Bases du système metrique décimal*, the scale of the enterprise left even the Emperor awed. 'Conquest is temporary,' he said modestly, 'this work will endure.'

Across the Channel the Ordnance Survey was in the midst

of its seventy-year project of triangulating Great Britain. In the Netherlands, Scandinavia, Russia and the Austro-Hungarian empire, governments had started or were about to embark on similarly exact and scientific surveys of their territory, using triangulation to calculate distance, celestial observation to establish location, and barometric pressure to indicate height. From the Arctic to the Mediterranean, from Ireland to the Urals, a series of geodetic surveys, precise enough to establish where each town and mountain stood on the earth's surface, were beginning to cover Europe.

By contrast, the maps of the United States were small-scale and produced by individuals like Thomas Hutchins. The only official, large-scale record was the series of surveyors' plats compiled at speed using only chain and compass, and their technical name, 'cadastral' or 'property registering', indicated their limited purpose. It was this lack that Jefferson and Gallatin intended the Coast Survey to fill. Whatever Congress might have had in mind, they wanted a scientific agency, the first created by the federal government, which would do for the United States what the Cassini family and Jean-Charles Borda had done for France, and William Roy and the Ordnance Survey were doing for Britain.

It seems clear that they also knew who they wanted in charge. Despite shortcomings in both language and temperament – towards the end of his life the New York newspaper *New World* referred to him as 'an old Swiss named Hassler who writes a miserable jargon which he calls English and scolds like a fish hag' – Hassler's proposal for the survey, written in French, was selected as the model to be followed.

The proposition that he put to Gallatin was to measure the Atlantic coast with a chain of enormous triangles whose sides would be sixty thousand to 100,000 feet long, resting on two precisely measured base-lines. In the 'miserable jargon' he called English, he emphasised the utter necessity of using only the finest instruments, including telescopes, theodolites, barometers, thermometers and chronometers, explaining that 'Good instruments are never to be found in shops, where only instruments of inferior quality are put up for sale. They must be made on command, and by the best mechanicians of London.'

The supremacy of British instrument-makers was universally acknowledged. At that very moment, Benjamin Latrobe, a surveyor with the Chesapeake and Delaware Canal Company, was writing to his employers about a telescope level he had bought from Mr Biggs, Philadelphia's finest maker of 'nautical, optical, geographical, surveying, gauging, gunnery, drafting, drawing, and leveling instruments'. It was, Latrobe commented sniffily, 'truly made for the American market as to filing, gilding, and engraving. It could not have been sold, I believe, in London.'

Only Britain had the metallurgists and machine tools, the glass-blowers, chemists and metal-workers, to produce instruments of the required accuracy. The most exacting craftsmen – 'artists', Hassler called them – were to be found in London, and the best of them was Edward Troughton. The German astronomer H.C. Schumacher once waited four years to get a zenith sector made by Troughton because no other was good enough. For the scientist Sir George Shuckburgh, Troughton constructed a yard with a built-in micrometer attached to a microscope so that the measurements marked in hairlines could be examined to an accuracy of one-thousandth of an inch. The Master of the Mint and General William Roy both ordered their definitive standard yards from Troughton; but Hassler paid him the finest compliment of all, naming his newborn son Edward Troughton Hassler.

Unfortunately for Jefferson's plans and Hassler's career, a trade embargo and then the War of 1812 with Great Britain delayed the acquisition of these essential instruments, and almost ten years passed before the Coast Survey began, by which time Jefferson had been succeeded as President by

James Madison. A small income from teaching, first at West Point then at Union College, a private school in Schenectady, New York, kept Hassler's family from starving, but it was not until July 1816 that he laid out the first two base-lines with infinite care, along the beach of what is now Coney Island and at Englewood, New Jersey. Despite the fineness of Hassler's instruments and the exactness with which he measured the azimuth of the Pole Star, the least educated chainman on the land survey would have laughed at the sight of his measuring chain. It had nothing to do with Gunter. Instead it consisted of eight links, each of which was exactly one metre long. Its length was measured against the metre that Hassler's friend Tralles obtained for him in 1799 while a member of the international metric commission.

After just one summer's work Congress closed down the survey for lack of money. Hassler retreated to Long Island, where he tried being the farmer he had originally wanted to be; but Marianne could not stand the isolation and ran away. Most of his earnings came from writing textbooks, and from a commission to examine the weights and measures used in United States customs-houses. It was not until 1833 that the Coast Survey was restored, with Hassler hired again as Super-intendent. He immediately celebrated with another London buying splurge – a new thirty-inch repeating theodolite built to his own design by Troughton, two microscopes from Dol-lond, and a Ramsden dividing engine so that in future the United States could make its own precision instruments. This time he used four iron bars, each two metres long, laid end to end to measure the base-line, and his triangles were so meticulously calculated that when they were checked in the 1970s the error rate was found to be just one in 100,000.

Next to prickly Rufus Putnam, the charmingly, Swissly pre-cise Hassler is one of the most attractive measurers of the United States. Nothing distracted him from his task. In sum-mer and winter, he dressed for simplicity's sake always in loose white flannel, merely subtracting or adding layers as the temperature rose or fell. Despite his strange accent and hopelessly unmilitary appearance, his West Point students adored him, and almost forty years later one of them, the redoubtable Colonel Joseph Swift, commander of the Corps of Engineers, wrote to Hassler's family recalling affectionately 'The intimacy between your father and myself . . . that remained unbroken through Mr. Hassler's undeserved vicissitudes to his death.'

To understand Ferdinand Rudolph Hassler, so odd in behaviour, so exact in performance, is to appreciate cuckoo-clocks. For his array of heavy brass instruments – the Troughton theodolite alone weighed three hundred pounds – he had a special carriage built with extra-wide wheels, heavily braced springs and cushioned boxes. It was painted canary yellow, and contained a suspended desk so that it could act as an office, and a folding bunk with a locker below holding wine, biscuits and cheese so that it could act as a home. When he grew short-sighted in old age, Hassler refused to be bothered with glasses but took to flicking snuff in his eyes – 'To excite the optic nerve,' he explained – so that as he bent over a map the stinging tears acted as a temporary contact lens. Nothing was allowed to distract him from the incessant drive for accuracy.

It was Hassler, for example, who discovered that most metre bars were minutely shorter than they should be, because while the ends were being filed off to bring them to the correct length the metal heated slightly and expanded. Once it cooled, what had been a metre contracted to something like six hundredths of a millimetre less than it should be. Hassler's concern for these tiny quantities was a symptom of his modernity. When Andrew Ellicott and Simeon de Witt, America's two leading surveyors of the eighteenth century, worked with him on defining the frontier with Canada, they found his approach frustrating. Writing to a friend, Ellicott complained that 'not more than one observation in ten can possibly be applied to the boundary – those that can are probably good, but their mode of calculation is laborious in the extreme'. But Hassler was not going to lower his standards for anyone. When at last he was allowed to begin work on the Coast Survey, he spent forty-three days measuring a base-line less than nine miles long.

This meticulous insistence on accuracy drove Congress to distraction, and messages winging from Washington to the yellow carriage would demand to know how long it would be before Hassler finished. In 1843 the *New World* declared that it was 'an evil hour when [Hassler was] selected as chief surveyor. He has been engaged in this for 13 [*sic*] years at the rate of 6000 dollars per annum, and as the results are not forthcoming Congress naturally wishes to know what he is about.' That winter he died, his triangles having reached no further than the southern border of New Jersey. But by then the exacting standards he set had become part of the United States Coast Survey. As a result, it took another fifty-five years to survey the entire coastline from Maine to New Orleans – a distance of 1623 miles, exactly – and every yard of it was measured in metres. Later the Coast Survey was extended to cover the entire United States, with the words 'and Geodetic' added to its title, and the whole landmass was mapped in the same careful, metric fashion.

There does not seem to have been any dissent at the time to Hassler's decision to use the metric system. The only measure specified by Congress was the league – it limited the survey's scope to 'within twenty leagues' of the coast – and he would not have been expected to use that archaic unit. His 1807 proposal for triangulating the coast expressed distance in thousands of feet. Yet there was never any question in his mind about the superiority of the metric system. It was not just more convenient, he once wrote, it 'produced results of considerable utility in many other respects, improvements in mathematical and natural sciences, and in mechanical arts, besides the establishment of the best and most consistent system of weights and measures hitherto devised'. For a scientific project like a geodetic survey, no other system was conceivable.

Jefferson, having left office before the survey began, does not appear to have been aware of Hassler's innovation. He never warmed to the metre, regarding the meridian basis as an unnecessary act of French nationalism which, as he explained in a letter to Patterson, forced other nations 'to take their measures from the standard prepared by France'. But his enthusiasm for decimals did not diminish – one of the gadgets of his old age was an odometer which divided the mile into hundredths, and he boasted that 'I find every one comprehends a distance readily when stated to them in miles & cents; so they would in feet and cents, pounds & cents, &c.' He had always supposed that once decimal measurement was adopted by 'the men of science', it would only be a matter of time before it was taken up by 'the tardy will of government who are always in their stock of information a century or two behind the intelligent part of mankind'.

That too was the view held by John Quincy Adams, the son of Jefferson's old antagonist and friend, the second President of the United States. In 1821 he delivered a report to Congress on the feasibility of 'establishing uniformity in weights and measures' which made it clear that whatever changes were made would happen very slowly indeed. He illustrated his point with a wry look at the success of the nation's decimal coinage in displacing the old British and Spanish currencies.

'Even now at the end of thirty years,' Adams wrote, 'ask a tradesman or shopkeeper in any of our cities what is a dime or a mille, and the chances are four in five that he will not understand your question. But go to New York and offer in payment the Spanish coin, the unit of the Spanish piece of eight [an eighth of a dollar], and the shop or market-man will take it for a *shilling*. Carry it to Boston or Richmond, and you shall be told it is not a shilling but nine pence. Bring it to Philadelphia, Baltimore or the City of Washington, and you shall find it recognised for an eleven-penny bit; and if you ask how that can be, you shall learn that, the dollar being [equivalent to] ninety pence, the eighth part of it is nearer to eleven than to any other number: and pursuing still further the arithmetic of popular denominations, you will find that half eleven is five, or at least that half the even-penny bit is the fi-penny bit, which fi-penny bit at Richmond shrinks to four pence half-penny, and at New York swells to six pence. And thus we have English denominations most absurdly and diversely applied to Spanish coins; while our own lawfully established dime and mille remain, to the great mass of the people, among the hidden mysteries of political economy – state secrets.'

Had he been able to look further ahead, Adams might not have been astonished by an official report made by a State Department representative to an international conference in Berlin in 1862, which revealed that almost eighty years after the decimal dollar became the currency of the United States, 'shopkeepers still more readily say two shillings and sixpence than thirty-seven and a half cents'. But even he might have been startled to find that in the twenty-first century, the quarter was still known as two bits, and that the New York Stock Exchange continued to divide it into eighths and sixteenths, as though Jefferson's decimal dollar were still the old Spanish kind.

Taking a long view, Adams was not inclined to recommend any urgent changes. The existing system was chaotic, according to his report, but he could not see any practical advantage in changing to a decimal system. 'A glance of the eye is sufficient to divide material substances into successive half, fourth, eighth and sixteenth . . . But divisions of fifth and tenth parts are among the most difficult that can be performed without the aid of calculation.'

Nor did he think that the metric system's scientific basis was especially relevant. 'It is of little consequence to the farmer who needs a measure for his corn, to the mechanic who builds a house, or to the townsman who buys a pound of meat or a bottle of wine to know that the weight or the measure which he employs was standardised by the circumference of the globe. Should the metre be substituted as the standard for our weights and measures instead of the foot and inch, the natural standard which every man carries with him in his own person would be taken away.' His conclusion was that of opponents of the metric system for centuries to come: 'The convenience of decimal arithmetic is in its nature merely a convenience of calculation.'

In the long term, Adams recommended that Congress should consult with other nations about co-operating with France to 'the final and universal establishment of her system', but meanwhile the best the United States could do was to wait until Britain, the economic giant of the age and the nation's closest trading partner, had established standards for its own weights and measures. Congress needed no other excuse to do nothing.

But Adams was wrong. Inactivity was no longer an option. The confusion of measures used in the thirteen original states had been carried by the public lands survey into ten new western states. On being admitted to the Union, most had legislated that standards of 'English measures' were to be used, without distinguishing between the different sizes of bushel, barrel and gallon. In Indiana and Illinois, for example, Virginian measures were used in the south of the state and New England units in the north. Vermont and the District of Columbia had inherited their own private confusions from their most immediate neighbours.

Louisiana, where fifty thousand French inhabitants had been converted by the Louisiana Purchase into Americans, faced the most contentious situation. They had always defined weights by the *livre* and *quintal*, and lengths by the *pied* and the *toise*; their rice was measured by the *boisseau*, their land by the *arpent carré*, and their cloth by the *aune*. Since 1814 it had been illegal to use any but American measures (in practice Georgia's), with fines of $50 for offenders, but when Monsieur Bouchon, Louisiana's Surveyor-General, wrote to Adams in 1820, he could report only mixed success. 'The ancient inhabitants are well enough satisfied with the American weights,' he noted, 'but all, and especially those in the country, find it very difficult to accustom themselves to the measures of length and [of area] and I think they will be long in doing so.'

This was a shrewd prediction. For years the characteristic shape of a Louisiana farm remained a narrow strip measuring two hundred square *arpents* (about 166 acres), and the resistance of farmers to the acre forced the land survey to abandon their squares around New Orleans in favour of the strip; to this day many Louisianans still refer to real estate in *arpents*. But as might have been expected, in New Orleans itself traders were more flexible, taking advantage of the change to buy goods by the *pied*, which was almost thirteen inches, and sell by the twelve-inch foot.

Confusion by itself was not enough, however, to break the Congressional repose. It was the trade between these western states and the east that finally forced them to act.

In 1825 Governor de Witt Clinton of New York opened the 350-mile-long Erie Canal linking the lake with the Hudson river, and predicted that, 'As an organ of communication between the Hudson, the Mississippi and the St Lawrence, the great lakes of north and west and their tributary rivers, it will create the greatest inland trade ever witnessed.' That turned out to be a modest forecast, for a ton of corn that had taken three weeks to be carried by wagon from Buffalo to New York at a cost of $100 could now be put on a barge and for just $6 would be delivered eight days later. Suddenly the produce of the mid-west that had been carried down the Mississippi or painfully by mule and wagon across the Alleghenies, gushed instead through the canal into the port of New York – millions of tons a year of grain, timber and coal.

It transformed New York into the premier market of the United States, and just as British tobacco-merchants had imposed their own definition of a hogshead on Virginia tobacco-growers, so New York brokers buying from mid-west farmers insisted on their own tons and bushels. But a New York bushel might contain as much as 2175 cubic inches of wheat or as little as 2104 cubic inches. The first figure applied to purchases, and the second to sales; multiplied by several hundred thousand tons a year, the disparity contributed significantly to the profitability of New York's trade.

Mid-west farmers suffered, but so did the United States's revenues, the greater part of which came from tariffs levied on goods passing through New York. Few customs-houses had standards of their own – in New York, federal customs officers depended entirely on the state's measures – and it was a poor trader who could not avoid paying at least some of his dues. The new states were above all farming states, whether they produced corn, wheat or cotton. It was in their interest not only to have uniform weights and measures, but to have revenue raised from tariffs rather than land taxes.

It was no coincidence that when the United States started to establish an official set of measures, the President was Tennessee-raised Andrew Jackson, the personification of the new western outlook. Nor did chance have anything to do with Jackson's choice of a commissioner to examine the state of measures held by United States customs-houses. There was only one person in the whole country whose obsession about accuracy to the last observable fraction of a measurement matched the task. In 1830 Ferdinand Rudolph Hassler was appointed to undertake the examination that was to lead the United States to a uniform system of weights and measures.

At the age of sixty, Hassler must have considered his life a failure. The Coast Survey had been stopped, his wife had left him, he had lost all his money, sold his survey instruments, his unique collection of metric and organic measures and his library of three thousand volumes, and survived by writing textbooks and eating cheap meals in Kinchy's Italian restaurant in New York. He had grown rather solitary, certainly a little cranky, although it was hard to tell whether this was a change or an exaggeration of something already present.

While they were together, he and Marianne had produced nine children, whom Hassler clearly loved. He made their clothes, baked bread and pastries for them, and as adults two of his sons chose to work with him. Nevertheless, as a husband he must have been exasperating, alternately absent-minded and stubborn about doing things exactly right. His capacity for concentration had always been extreme, making him, as one member of his class at Union College recalled, a hopeless teacher: 'He became so absorbed in demonstrations, his students would walk out on him, and he would never be conscious of their absence.' They made up a story about him reading a letter and being so lost in its contents that he tripped over and fell, yet continued reading while he lay on the ground until he had finished it. To those who knew Hassler it sounded credible.

Despite his strange appearance – the red-rimmed, snuff-inflamed eyes, the white suit stained with wine and crumbly with biscuits – Hassler was not a freak. In his single-minded passion for accuracy, he was simply a man out of his time. A generation later, science had adapted itself to the supreme importance of precision, writing up every detail of how experiments were conducted, and recording results to three and four places of decimals. In 1832, when Hassler produced his report on the customs-house measures for the President, he still had to make the case for exactness, and in a passage as remarkable for its passion as its fractured English he delivered what was in effect his testament of faith: 'A system of weights and measures must be a careful scientific operation in which the accuracy aimed at must far exceed that required for practical use . . . What is done in such a work is done for the future, the improvement of science always spreading more in common life; if such [a standard of accuracy] is not ahead of its time, even if possible of the science of it, it very soon drops back, behind even the wants of the nicer social intercourse; and such an epoch approaches always more rapidly with the greater means of science.' Which meant, roughly, that science brought progress, that the better the science the faster the progress, but that progress must inevitably consume the science that made it possible.

Yet Hassler did not value accuracy for its own sake, or even for science's, but, as the opening words of his report explained, because 'Among the means of distributing justice in a country, which is the aim of the establishment of Governments, rank unavoidably the fixation and distribution of accurate weights and measures for all the daily dealings of active life.'

Although he had introduced the metric system to the Coast Survey, there was never any possibility that Hassler would attempt to do the same for the United States as a whole. But as a believer in the system, he could not help expressing his regret that Jefferson had failed in his efforts to link reform of weights and measures to his design for creating states out of the empty west. 'It would undoubtedly have been a great advantage to the country,' he wrote, 'had a regular system, founded upon a scientific base and single unit, been established together with the first regulations for the organization of the country, before the great increase of population, and consequent great active intercourse had created and increased the difficulty of attacking old habits.'

That moment had passed. In 1790 there had been fewer than five million inhabitants to convince of the need for change; in 1830 there were almost thirteen million, nearly all of them used to traditional, four-based measures. Hassler's task was simply to establish standards of uniformity, and some of the work had already been done for him.

In 1824, the British government had at last replaced all the conflicting measures that so upset the Carysfort Committee eighty-five years earlier and established a single system not only for itself but for the Empire. The Imperial weights and measures were based on three fundamental standards: Bird's 1760 yard, which itself was derived from Elizabeth I's 1588 measure; Bird's 1758 Troy pound, which was taken from the gold merchants' and apothecaries' measure made legal for the first time in Henry VII's 1496 statute; and a single gallon size for both liquid and dry measures, defined as being large enough to contain ten pounds avoirdupois of distilled water. The first two were utterly traditional, but the gallon (equivalent to 277.421 cubic inches) was an innovation born from the committee's inability, even after ten years of intellectual bloodshed, to agree on one definitive version out of all the different, contradictory ale gallons, wine gallons and corn gallons already on the statute book.

Having battled their way to agreement on the standards, the members of the measures committee saw their work destroyed in 1834 when the Houses of Parliament, where the Imperial yard and pound were kept, caught fire, and in the flames the two meticulously engineered originals were reduced to iron puddles. Another ten years passed before replicas of sufficient accuracy could be constructed, but from these two units were derived all the others, from grains to tons, and from inches to miles. Hundreds of standards were sent out to local authorities, and magistrates were ordered to enforce their use.

One of the units, the Troy pound, had already become an unofficial standard for the United States after a brass copy was purchased by Albert Gallatin, while ambassador in London, and presented to the Philadelphia Mint in 1828. It remained the official standard for measuring the weight of United States coins until 1911. Since an avoirdupois pound could be derived from the Troy variety, this smooth, circular brass weight served for a time as the absolute standard for all American weights. The presumption was that Hassler would take other Imperial units to serve as his standards.

His report took over a year to compile. Many of his experiments were carried out in winter so that instruments could be tested above and below freezing point. For standards he used the Troy pound in the Mint and the splendid brass eighty-two-inch Troughton measure he had had made for the Coast Survey in 1814. The yard and every gradation from nought to eighty inches were cut into it. But to satisfy Hassler's lust for exactness, John Vaughan, who had bought the Tralles metre bar and kilogram so that Hassler had enough money to eat, lent them back to serve as additional checks.

In a typical passage in his report, Hassler explained the difficulty of comparing a customs-house weight with one of his standards. Each had to be placed in a scale whose point of balance was for utmost sensitivity a knife-edge, but its accuracy 'depends on the sharpness of the knife edges, and these must lose their edge under a heavy pressure; thence besides their very expensive construction, they require the constant presence of an artist to repair them after a few weighing operations'. There was always a price to be paid for Hassler's work, but no one ever doubted that the knife-edge of his scales would cut like a razor. When it proved impossible to maintain that sharpness, he built a hydrometer in which weight was determined by the displacement of water.

Hassler's report revealed what the Treasury Secretary Louis McLane called 'a serious evil', although he was referring to loss of revenue rather than the injustices that resulted from different measures being used, or in some cases being totally absent, in ports from Boston to New Orleans. 'It is believed, however,' McLane continued, in a sentence that was to signal the breaking of the logjam, 'that this department has full authority to correct the evil, by causing uniform and accurate weights and measures, and authentic standards; to be supplied to all custom-houses.'

They were to be made at the United States Arsenal in Washington, 'under the immediate personal superintendence of Mr. Hassler . . . with all the exactness that the present advanced state of science and the arts will afford'. If that rang like the sound of trumpets in Hassler's ears, it must have seemed as though the full orchestra had struck up in August of that year, when the Coast Survey was re-instituted. For the last ten years of his life, he remained in charge of both, working on the survey in summer and the weights in winter.

In constructing the standards, Hassler displayed the full range of his gifts for exactness. Copies of weights from around the world were ordered, alloys of metal were assayed, then weighed in a vacuum and at normal sea-level pressure, at freezing and at sixty-two degrees Fahrenheit. But that was only the start. Eventually each standard had to be weighed on precisely calibrated scales. Because the tiniest movement of air would affect the scales, Hassler covered up the windows of the laboratory to prevent draughts and to stop the sun's rays heating the room and creating temperature currents. 'I took besides the precaution to nail to the side of the bench directed towards the windows, papers that reached several inches above it to prevent any draft over the scale,' he reported, 'and other large whole sheets [were nailed] on the side towards the observer, near the microscopes, so as to intersect the communication of the heat of the body of the observer to the microscopes and scale, or the effect of his breathing.' Having set up the laboratory, Hassler would leave everything in place until the following day so that the disturbed air and temperature had an opportunity to settle down. On his return he wore gloves so that when he placed the weights on the scales no heat would be transferred from his hands.

As reports on Hassler's construction of the customs-house standards reached Congress, his extraordinary commitment to the uttermost achievable degree of accuracy began to affect the way Congressmen thought about his goal. It was still pictured more in terms of revenue than of justice, and Hassler was regarded rather with irritation than respect, but for the first time since George Washington had reminded Congress of their constitutional responsibility for weights and measures, the subject was taken seriously. On 14 June 1836, a joint resolution of the Senate and the House of Representatives ordered that 'a complete set of all weights and measures adopted as standards [for customs-houses] . . . be delivered to the governor of each State in the Union . . . to the end that a uniform standard of weights and measures may be established throughout the United States'.

What was odd about all this, and remains astonishing, was that there had never been any formal definition of what those weights and measures should be. It was simply left to the Secretary of the Treasury, who took his lead from Hassler – if it was good enough for him, it was good enough for the United States. And so from making standards for the customs-houses, Hassler went on to make a set for every state in the Union. It was not until 1857 that the weights and measures he made were formally made legal throughout the land as the American Customary System of Weights of Measures.

The yard was specified as the distance between the twenty-seventh and the sixty-third inch engraved on a silver scale inlaid down the centre of Troughton's eighty-two-inch scale, which is still preserved in the museum of the National Institute of Standards and Technology in Washington. The basic weight was the Troy pound in the Philadelphia Mint, which was scaled up to give the avoirdupois pound by the ratio of seven thousand grains to 5760 grains, the actual standard being a brass weight marked with a star, and consequently known as the 'star' pound. Both of these were the same as the Imperial units, but instead of selecting the new all-in-one British gallon, the United States Treasury decided that the distinction between liquid and dry measures should be kept, initially for customs duties and then for the standards themselves. From the grab-bag of old definitions, Hassler's elaborate comparisons of different measures determined the selection of Queen Anne's wine gallon of 231 cubic inches for liquids, and for grains the Winchester bushel of 2,150.42 cubic inches, each of which was smaller by about 17 per cent and 3 per cent respectively than the new Imperial gallon.

For convenience, he recommended no standard for the British stone, weighing fourteen pounds, which quickly dropped out of use in the United States, and decided to clean up the incongruity of a hundredweight weighing 112 pounds. In the United States it would weigh one hundred pounds, and the United States ton would come in at two thousand pounds rather than the British 2240 pounds. It was not quite decimal, but it was a gesture.

Since the pound was Roman in origin, and the yard Saxon, while the wine gallon was first recorded in Magna Carta, and the Winchester measures could be traced back to King Edgar in the tenth century, the set of weights and measures that emerged from Hassler's workshops was truly traditional. In science he might prefer the stark simplicity of the metre, but for practical purposes he recognised that the choice 'of a set of standards in general depends upon the individual use made of them', and for the United States that meant the old four-based measures. It was his unique achievement to have established both systems for use in his adopted country.

In November 1843 Hassler fell heavily while trying to protect one of his precious surveying instruments in a storm, and a few days later caught pneumonia and died. As often happens, all the stories of his exasperating behaviour instantly became affectionate reminiscences. One told by President Jackson was of Hassler coming to demand that his salary for the Coastal

Survey be raised to $6000 a year. When Jackson objected that this was as much as Hassler's boss, the Secretary of the Treasury, was being paid, Hassler exclaimed in his uncertain but utterly truthful English, 'Plenty Mr Everybodys for Secretary of ze Treasury, but – only vone Mr Hassler for ze Head of ze Coast Survey!' Recognising that in this, as in everything, Hassler's accuracy could not be questioned, Jackson gave him the raise.

SIXTEEN

❊

The Dispossessed

IT WAS ESTIMATED THAT seventy thousand people passed through the Crystal Palace in south London on 1 May 1851. Everything about Sir Joseph Paxton's structure was fabulous – walls and arched ceilings of clear glass prefabricated round an intricate skeleton of slender cast-iron rods soared over a hundred feet in height, stretched more than six hundred yards in length and covered an area of twenty-three acres. Yet it was the gigantic crowds, the largest most of them had ever seen under one roof, that witnesses repeatedly commented on. The Great Exhibition for which the palace had been constructed had on display more than fourteen thousand inventions from around the world, ranging from machine-tools to pins, and at the opening ceremony, when Queen Victoria appeared to the sound of a gigantic organ accompanied by a massed choir thundering out the Hallelujah Chorus from Handel's *Messiah*, she was followed by an international train of princes, heads of state and ambassadors. The sight almost overwhelmed William A. Burt, a surveyor from Michigan, who was there.

'Dear companion,' he wrote that evening to his wife Phebe, back in Michigan, 'This grand precession consisted of nearly every nation under heven in one harmonious band engaged in one object; the like was never seen before. I cannot give you anything like an inteligent view of it. It exceeds by far anything and everything that I had imagined. There is here everything to be seen in nature & art that this world produces . . .'.

Square-shouldered and bearded, Burt had spent the greater part of his life in the emptiest parts of the frontier among the mountains and lakes of Michigan and Wisconsin, and nothing had prepared him for this experience. It was his invention of a solar compass that brought him across the Atlantic. In the Michigan peninsula, enormous iron deposits threw out the land surveyors' magnetic compasses so badly the needles danced back and forth through as much as sixty degrees. What Burt had devised was an instrument based on the sextant which used the sun to give true north, and without it measuring out straight lines would have been impossible. Its value had led it to be chosen as one of 560 American exhibits, alongside Cyrus McCormick's mechanical reaper and Samuel Colt's repeating revolver, on display at the Great Exhibition.

The exhibition, however, was dominated by the display of industrial muscle that powered the British Empire. Its star was Sir Joseph Whitworth, whose engines cut and ground out machine parts to a tolerance of one ten thousandth of an inch. A generation earlier, in 1824, when the standards for the Imperial measures were created, that degree of precision had represented the very limit of scientific accuracy; now it was a commercial commonplace. It was as Hassler had foreseen – accuracy's children were devouring their parents, the finer measurements driving out the rougher. By 1851, Whitworth's measurements, and particularly Whitworth's screws, so dominated metal-cutting technology that they virtually held the Industrial Revolution together.

By that date too it was clear that in Britain, unlike the United States, industry had overtaken land as the prime source of wealth creation. But to William Burt everything about the two countries was different. 'I had a fine view of this ancient country,' he told Phebe after his train journey from Liverpool to London, 'it is all cultivated like a garden, it is unlike any thing that I have seen in America.' Above all there was the social divide, and in his next letter William's pride at finding himself in royal and aristocratic company struggled with the democratic instincts of a mid-westerner. 'I have had an opportunity to associate with Lords and Noblemen and to be in the presence of Queen Victoria and Prince Albert and other Princes, for whome mutch greate veneration is had in England, *as if the gods were present*,' he confided to Phebe, 'but I could not feel any such veneration for them, for I saw and felt that they were human beings, but a worm of the dust like myself.'

Thomas Jefferson would have approved these sentiments and, perhaps rightly, given the credit to his own scheme of democratic land distribution. But there was much more in common between the hierarchical British Empire and the republican United States than he or Burt might have cared to admit.

By the Treaty of Guadalupe–Hidalgo, signed in 1848 with its army encamped in Mexico City, the United States forced the Mexican republic to surrender a territory that included present-day California, Nevada, Utah, most of Arizona, New Mexico and Colorado and part of Wyoming. Added to the annexation of Texas in 1845, this constituted an area larger than the Louisiana Purchase; an understandably complacent census official pointed out that the addition made the United States 'of equal extent with the Roman Empire or that of Alexander, neither of which is said to have exceeded 3,000,000 square miles'.

Thanks to the grid, this territory was being colonised at phenomenal speed. Had he not been in London, William Burt would have been on the other side of the world, on the Oregon coast, where his friend William Ives was running the Willamette meridian north from a point near Portland, using a Burt solar compass. Just sixty-six years after Thomas Hutchins'

chainmen took their first westward steps on the banks of the
Ohio river, the land survey had reached the Pacific, and
already the land office of Oregon had registered a claim from
John Potter, a married man of Linn County in Oregon Territory, for '320 acres of Land, known and designated in the
Surveys and Plats of the United States as Part of Sections 22
& 27 T[ownship] 9 S[outh]. R[ange] 2 E[ast]'. Although most
of the country from the Mississippi to the west coast remained
unsurveyed, the squares now spanned the continent.

The only other imperial power to rival this rate of territorial
acquisition during the nineteenth century was Britain. Following the loss of its North American colonies, it had conquered
India, occupied Australia and New Zealand, pushed its Canadian claims to the Pacific, moved into southern Africa and
established a string of colonies reaching to the borders of
Ethiopia. It was the need for uniform weights and measures
throughout its growing number of possessions that impelled
the British government to pass the 1824 Imperial Weights and
Measures Act.

Britain and the United States shared other characteristics –
language, elected government, and a legal framework based
on the common law – and they were unique in their greed
for territory, their respect for property, and their faith in technology. And both measured out the wilderness with Gunter's
chain. The manner in which land passed into private ownership in the British Empire, however, was largely determined
by Edward Gibbon Wakefield, who combined the qualities
of Moses and Svengali, leading thousands to their promised
lands, and manipulating politicians as easily as he did the two
teenage heiresses he kidnapped and married. His dominating
personality was said to come from a secret use of hypnotism,
and one political opponent declared that the only way to resist
him was to hate him intensely. In 1849 Wakefield published
A View of the Art of Colonization, which described the programme
that he had persuaded the colonial authorities to adopt in

Australia, New Zealand, South Africa and parts of Canada. The inspiration for it had come to him while he was in prison after the second kidnapping.

The Wakefield plan was first tried out in Australia in the 1830s. In place of the free distribution of what was known as 'waste land' to any settler who claimed it, usually in the form of leaseholds or with quit-rents to emphasise the transaction's feudal nature, Wakefield had argued that land should be sold outright for a 'sufficient price'. This was a sum large enough to achieve two objects: to build up a fund which would pay for more immigrants to come free of charge to the colonies, and to make it difficult for the new arrivals to buy land until they had worked for three or four years. In Australia a sufficient price for a hundred-acre parcel of land was deemed to be £1 (about $5) an acre in 1842, because £100 represented the cost of sending out the four labourers needed to make the parcel productive. The effect was to create a pool of labour for existing farmers; and when the new immigrants had saved enough to buy their land, they would have a capital asset whose value could be enhanced by the work of new arrivals.

Wakefield was as much a social engineer as Thomas Jefferson, believing as strongly as he that social virtues arose from owning land – but his perspective was Hamiltonian. The society he envisaged was structured, enabling surpluses to be accumulated and leisure enjoyed. So long as land was given away, he argued, its owners would be too poor to do more than scrape a living, leaving 'money-getting the universal object; taste, science, morals, manners, abstract politics neglected'.

What Wakefield helped foster was a sense of structure – that land ownership was open to anyone, but that it had to be earned by service to others – which fitted naturally into the social pyramid created by the Empire. Particular circumstances affected the way landholding developed – the existence of industrialised Britain as a market for food, for example, encouraged pastoral farming with large acreages for sheep and cattle – but the overall context was provided by Wakefield's programme. Its influence was felt around the world.

In Australia, the practice of free leasehold grants ceased in 1831, and the transportation of convicts to New South Wales, once the only supply of cheap labour, was halted within a few years of assisted immigration beginning. Graziers searching for land to feed their animals had once simply squatted on ground beyond the authorities' control. Now, as the land acquired value, they registered claims to gigantic spreads, making them the forerunners of the wealthy 'squattocracy' occupying thousands of acres that nineteenth-century Australians loved to hate. Within a generation their example was being followed by an influx of owner-pastoralists who triggered the first great land boom in Australia in the 1860s.

In Canada, the start of Wakefield's scheme in 1841 marked a break with the different feudal hierarchies created by the French seigneurial system in Quebec or Lower Canada, and by the arbitrariness of British land allocation in Ontario or Upper Canada. That the character of the united Canada, also born that year, grew increasingly democratic and increasingly capitalist was due not simply to its new Constitution, but to an increasingly vigorous market in property. By the time the provinces of Alberta, Saskatchewan and western Manitoba were opened up in the late nineteenth and early twentieth centuries, Canada was ready to measure them out in the free enterprise fashion operating below the forty-ninth parallel, with the land squared off into 640-acre sections subdivided into halves and quarters.

The programme worked because of the legal presumption, broken in the United States in 1776, that colonised land remained 'Crown land' that could be allocated in whatever fashion the government thought best. One consequence was that in Canada native inhabitants could cite George III's proclamation guaranteeing them the inviolacy of territory west of the Appalachians, and so claim some protection against the incomers. Compared to the uncontrolled rush for land that developed in the United States, this had the advantage of encouraging a relatively ordered pattern of settlement, and although in most countries surveys were metes and bounds, the chaos of land distribution experienced by the Southern states of America was largely avoided. It operated most smoothly in New Zealand, where, almost uniquely, the indigenous people, the Maoris, retained some of their land.

Yet in the end, wherever the nineteenth-century surveyors unrolled Gunter's chain and drew maps recognised by English common law as a record of a property claim, they achieved the same result. Only in North Africa, where France transformed Arab land into French estate with the help of *géomètristes* using the metric system, was there anything comparable to the spread of the Anglo-American property-makers in the nineteenth century. Everywhere it depended on being able to measure and describe the land more precisely than the indigenous peoples could. Where the first inhabitants could meet those demands, their rights might be respected. Where they could not, their way of life was extinguished – and what happened to Australian Aboriginals, New Zealand Maoris, native Zimbabweans and southern African Xhosas bore a marked resemblance to the fate of American Indians.

The territory that William Burt surveyed was originally Ojibwe or Chippewa land. Barely a decade before Burt began to run the Michigan principal meridian due north to the Upper peninsula, Henry Schoolcraft, the Indian Agent for the Great Lakes region, had persuaded the Ojibwe to sign a treaty transferring the title of their land in northern Michigan and Wisconsin to the United States.

Given the difficulty of running a straight line through drifts of winter snow, and clouds of summer mosquitoes, and depths of permanent swamp, Burt could have been forgiven for wondering, as other surveyors did, why Schoolcraft bothered. One page of Burt's field notes records him entering a marsh at 5.68 chains from his starting point. At forty chains, where he was supposed to insert a post, his note runs: 'No post set nor bearing taken. Water 3 feet deep.' When he did emerge onto dry land at 71.20 chains from his starting point, it was to plunge into prickly undergrowth that tore his clothes to shreds.

'Dear Companion,' he wrote to Phebe, 'I am now more than halfway to Lake Superior from Mackinaw, and about 40 or 50 miles from any Settlement in the midst of a swamp about twelve miles in diameter but expect to get out tomorrow as I can see high Beech and maple Land to the North and a River between [us] and it, but I have just made and launched a Canoe. Nothing that will be news to write, but wish to hear from home. My Coat and Pantiloons are most gone. If you could make me a frock [long coat] like that of Austin's [their son] and a pair of Pantiloons of the strongest kind of Bedticking they would I think stand the Brush.'

Since no squatters had arrived in Michigan before them, some surveyors took advantage of the general ignorance to complete imaginary plats and field notes. One whose crimes were discovered pleaded reasonably enough that the deceit was not important 'because the land would not be sold in ten centuries'. William Burt was not that kind. His conscientiousness can be guessed from a detail. He discovered that in the bitter frost of a Michigan winter, Gunter's chain would contract by as much as an eighth of an inch. Compared with the errors that arose from the unpredictable swings of the compass and the viciousness of nature, this tiny alteration might have been overlooked; but Burt made it his habit to build a fire each morning so that he could warm up his chain to its full twenty-two yards.

Driving through the Michigan peninsula today, it takes an effort of imagination to see past the soya fields and groves of black walnut to that raw land beneath; but the north–south, east–west roads that lead seamlessly from neat farms into the neon sprawl of shopping malls and condo developments explain how the transformation came about. Below the roads run the surveyors' lines which squared off the wilderness, made it ready for sale, provided a source of income for schools and a university, and constructed a shape for county and state government. At $3 a mile it was one of the great bargains of the century.

But one other cost should be thrown in. Burt's meridian and the surveyor's squares obliterated something else, and ironically it is due as much to Henry Schoolcraft as anyone else that its value is known today. In a series of monumental studies of Ojibwe culture, Schoolcraft presented a detailed account of a society with an intimate, animistic connection between the human spirit and the natural world. In 1855, Henry Wadsworth Longfellow published his own version of that society's legends in an epic he called *The Song of Hiawatha*.

> Bright above him shone the heavens,
> Level spread the lake before him;
> From its bosom leaped the sturgeon,
> Sparkling, flashing in the sunshine;
> On its margin the great forest
> Stood reflected in the water,
> Every tree-top had its shadow,
> Motionless beneath the water.

Before the words were published, the survey had marched across Hiawatha's forests, revealing much of it to be commercially valuable white-pine that the lumber industry would fell – and in Negaunee on the Upper peninsula beside one of Hiawatha's lakes Burt had discovered the richest lode of iron ore yet found in the United States, which would be mined until the tailings turned the lake orange. On either side of long, arrow-straight roads in northern Michigan and

Wisconsin, billboards today alternately advertise Ojibwe casinos and the Gamblers' Anonymous warning, 'If you are gambling more than you can afford, you may have a problem.' It seems like a kind of revenge on the lines beneath.

It was Thomas Jefferson's belief – which, like so many others, became government policy – that as 'prime occupant' in the continent, the United States had 'the exclusive privilege of acquiring the native [title to land] by purchase or other just means. This is called the right of preemption . . . There are but two means of acquiring the native title. First, war; for even war may, sometimes, give a just title. Second, contracts or treaty.' In the years after the defeated nations of the Western Confederacy were summoned to Fort Greenville in 1795, Indian treaties became almost routine. So certain was the federal government of the eventual outcome that the 1804 Land Act extended the Surveyor-General's power 'over all the public lands of the United States to which Indian title has been or shall hereafter be extinguished'.

Every Indian war from Fallen Timbers in 1794 to the final massacre at Wounded Knee in 1890 had its origin in the hunger for land, and nearly every treaty that followed involved the transfer of ownership of land. Because US law was equivocal about the transfer of title where coercion was involved, the policy was to accompany victory with purchase, generally in the form of money, gifts and the guarantee of an alternative home with well-defined borders. The wording of the treaty would make it clear that the contract was voluntarily entered into, and for reasons other than defeat. Thus, the Kaskaskia were said to be ready to sell their territory beside the Mississippi because they were 'reduced by the wars and wants of savage life to a few individuals unable to defend themselves against the neighboring tribes'. More land up to the Wabash was ceded by the Delaware because, Jefferson asserted, they desired 'to extinguish in their people the spirit of hunting, and to convert superfluous lands into the means of improving what they retain'. In the south, the Choctaw decided to sell their land, 'being indebted to their merchants beyond what could be discharged by the ordinary proceeds of their huntings'. Their Chickasaw neighbours gave up land in Tennessee and Mississippi because, according to the 1832 treaty, they 'find themselves oppressed in their present situation; by being made subject to the laws of the States in which they reside'.

These reasons were no doubt in part genuine, and a price was certainly paid, but they camouflaged the reality that the United States was making an offer that could not be refused. The treatment of the Cherokee made that starkly clear. Having ceded over three million acres in what is now Kentucky and Tennessee by a series of treaties, the nation eventually split into two. In 1817 the western, more traditional half agreed to move west to new lands in Arkansas where they could follow their old ways, only to be forced out ten years later by the arrival of more settlers, and sent further west to Indian Territory in what is now Oklahoma. The eastern Cherokees chose to become Americanised in order to keep what remained of the homeland that had been guaranteed to them by the earlier treaties. Under their half-Scots chief, John Ross, they built schools, elected a ruling council, wrote a constitution, and farmed as successfully as any other Americans. By 1828 they were a prosperous community, but their land was wanted by Georgia speculators.

Despite two Supreme Court decisions affirming the Cherokees' right to the land, the federal government under President Andrew Jackson, who first made his name in campaigns against the Creek and Seminole peoples in Florida, forced them to accept yet another treaty, surrendering it in exchange for $5 million and seven million acres in Oklahoma. When they delayed moving out, they were driven off by the army, and in the course of their pitiful journey to Oklahoma in the winter of 1838, better known as the Trail of Tears, some four thousand of the sixteen thousand who started out died from disease, hunger and bitter cold. The implication was clear – whether the Indians moved or stayed, accommodated or resisted, the title to their land had to be transferred.

The pattern rarely altered: incursion by small groups of settlers, growing tension, Indian violence, American retaliation and the intervention of the United States Army. In Jefferson's presidency alone, thirty-two treaties extinguished Indian titles to most of the land east of the Mississippi, and the missionary John Heckewelder described in his memoirs the relentless pressure on the Native Americans: 'They say that when they had ceded lands to the white people, and boundary lines had been established – "firmly established!" – beyond which no whites were to settle, scarcely was the treaty signed, when intruders again were settling and hunting on their lands!' When the Indians complained, 'the government gave them many fair promises, and assured them that men would be sent to remove the intruders by force from the usurped lands. The men indeed came, but with chain and compass in their hands, taking surveys of the tracts of good land, which the intruders, from their knowledge of the country had pointed out to them.'

With a mixture of desperation and grim humour, the Seneca chief Red Jacket asked for some breathing space in a famous speech in 1829, much quoted by contemporary newspapers: 'Brothers, as soon as the war with Great Britain was over, the United States began to part the Indians' land among themselves . . . permit me to kneel down and beseech you to let us remain on our own land – have a little patience – the Great Spirit is removing us out of your way very fast; wait yet a little while and we shall all be dead! Then you can get the Indians' land for nothing, – nobody will be here to dispute it with you.'

With the passage of the Indian Removal Act in 1830 it became government policy to relocate Indians to land west of the Mississippi. In the north the Shawnee, Wyandot and Delaware nations were moved west from the Lake Erie region to what is now Nebraska and Kansas. In the south, Creek, Choctaw, Chickasaw and Seminole columns followed the Cherokees to Arkansas and Oklahoma.

In 1851 on the Pacific coast, Burt's assistant, William Ives, began to run the Willamette meridian from Portland, Oregon, north to the Canadian border. Eventually every acre in Oregon and Washington state would be related to it, including the rich coastland along the Puget Sound. Some of the land acquired by the United States had belonged to a small tribal group, the Duwanish-Suquamish, who were relocated to the Sierras. The protest of their chief, Sealth, was imaginatively reconstructed thirty years later by an onlooker, Henry Smith, and in the twentieth century was further embellished to become a well-known environmental protest: 'We do not own the freshness of the air or the sparkle of the water. How can you buy them from us?' Beneath the elaborations, however, Smith's words clearly convey not just the intense bond that Seattle (the American version of his name) felt with the land, but Smith's discomfort at what was entailed in America's 'manifest destiny' to fill the continent coast to coast.

'The Indian's night promises to be dark,' he wrote in Sealth's name. 'No bright star hovers about the horizon. Sad-voiced winds moan in the distance. Some grim Nemesis of our race is on the red man's trail, and wherever he goes he will still hear the sure approaching footsteps of the fell destroyer and prepare to meet his doom, as does the wounded doe that hears the approaching footsteps of the hunter. A few more moons, a few more winters, and not one of all the mighty hosts that once filled this broad land or that now roam in fragmentary bands through these vast solitudes will remain to weep over the tombs of a people once as powerful and as hopeful as your own.'

By the second half of the century, Smith's ambivalence was widely shared. It was not only Longfellow's *Hiawatha* that found a readership attracted to its picture of the American

Indians' pristine world. Many journalists gave sympathetic accounts of the words and demeanour of the people who were losing their land. Artists depicted them in noble pose – even the Bureau of Indian Affairs, whose purpose was to move them off their land, covered its walls with portraits of the noble red man.

When the Willamette meridian pushed the Nez Percé from their land in Oregon, and their leader, Chief Joseph, mounted a long, bloody defence against the United States cavalry that ended in 1877 in his capture on Bear Paw mountain, it was the United States press that published his eloquent explanation of why he had resisted. Sometimes their versions sound embellished – 'Tell your people that since the Great Father promised that we should never be removed, we have been moved five times. I think you had better put the Indians on wheels and you can run them about wherever you wish' – but other accounts caught the authentic identification with a land where his forebears had lived and died, and where his father was buried. 'I love that land more than all the rest of the world,' said Chief Joseph. 'A man who would not love his father's grave is worse than a wild animal.'

But the ambivalence did not affect the outcome. Early in Jefferson's presidency, when settlers were beginning to push into Illinois, one of the native inhabitants was presented to him. 'I am a Kickapoo,' explained Little Doe, 'and drink the waters of the Wabash and the Mississippi.' But the third principal meridian ran through there, one of the squares became Township 14 North, Range 5 East, also known as Sugar Creek, Sangamon County, Illinois, and the Kickapoo were moved to Oklahoma. Seventy years later, despite Chief Joseph's explanation that it was the land around his father's grave that made him who he was, a Nez Percé rather than a wild animal, the Williamette meridian moved him out of the Sierras just as effectively as the Third Principal had shifted Little Doe, and sent him and his people down to join the Kickapoo in Oklahoma.

In the 1880s even the Indian Territory in Oklahoma came under pressure from Texan and Kansan settlers. To forestall the threat of bloodshed from squatting by the Boomers, as they were called, the federal government surveyed two million acres of the Territory and declared that they were to be given free to the first people to claim them. On 22 April 1889, about fifty thousand would-be claimants crowded up behind a line drawn by the army near the railroad track. At noon a gun fired, and they raced forward to stake their claim, on horseback, on wagons, on foot, even on bicycles. Hamilton Wicks, who was on foot, remembered the wild scramble up the hillside with thousands on either side making for the same area. 'The race was not over when you reached the particular lot you were content to select for your possession. The contest was still who would drive their stakes in first, who would erect their little tents soonest, and then who would quickest build a little wooden shanty. The situation was so peculiar ... It reminded me of playing blind man's buff. One did not know how far to go before stopping, it was hard to tell when it was best to stop and whether to turn to the right hand or the left. Everyone appeared dazed.'

Before the gun fired, the land over which the settlers swarmed was Indian territory which had been surveyed and laid out into sections, halves, quarters, even quarter-quarters. Now it was property belonging to the claimants. The entire century was compressed into those frantic minutes.

SEVENTEEN

The Limit of Enclosure

W HAT HAPPENED TO Johann August Sutter might serve as the definitive example of the importance of measuring land in order to make it property. He was Swiss, a fact which worked in his favour when he arrived in San Francisco in 1838 in search of land. The Mexican authorities wanted settlers to occupy its empty territory. In twenty years they made 813 large land grants in California, compared with just twenty-five made by the Spanish in the previous half-century; but as in Texas the most vigorous applications for Californian land came from Americans, whose country posed the greatest threat to Mexican possessions. 'Hordes of Yankee immigrants,' complained the Governor of California, 'are cultivating farms, establishing vineyards, erecting mills, sawing up lumber, building workshops and doing a thousand other things which seem natural to them.'

Sutter's request for land accordingly fell on receptive ears, and he was granted eleven square leagues wherever he chose to settle. He selected the valley where the Sacramento river met the American river, low-lying land but flat and very fertile. He drained and fenced, ran cattle and horses, built houses, a tannery, a mill, and planted a vineyard. And another grateful Governor awarded him a further thirty-three square leagues

The document accompanying the grant only noted the main boundary markers, and the plat or *diseño* was no more than a sketch map. In Mexico, ownership was created not just by the paperwork, but by occupation and communal acceptance. The telling evidence in determining whether this meant modern property was revealed, as always, by the measurements.

'There is no long measure corresponding to our acres,' commented Josiah Gregg when he visited northern Mexico in the 1840s. 'Husbandmen rate their fields by the amount of wheat necessary to sow them: and thus speak of *fanega* land – a *fanega* being a measure of about two bushels – meaning an extent which two bushels of wheat will suffice to sow.' The unit was, therefore, variable, and a measure of yield rather than area. Larger grants of land were measured by the square *ligua*, which, under the Spanish, also varied depending on whether it referred to the area of a *pueblo* (a village) or a *rancho* (an estate). Nevertheless change was already underway by 1821 when Mexico achieved its independence, and the square *ligua* became more or less standardised at twenty-five thousand square *varas*, or 4428 acres. It was significant, however, that when Stephen Austin received a grant of Texas land from the Spanish and Mexican governments in the 1820s, he decided to survey it and specify that the *vara* measured precisely 33⅓ inches. In American eyes, ownership required exactness.

Sutter failed to take that precaution, and probably did not feel the need for it. He was master of about a million acres, and lived like a feudal baron, laying out a town, New Helvetia, now Sacramento, guarded by his own palisaded fort, Fort Sutter, where he provided lavish hospitality for American migrants coming down from the Oregon Trail. Up to eight hundred Californian Indians whose land he had taken provided slave labour and a private army.

By the 1840s Sutter had made enough money to buy the Russian claim to northern California – measured in *versts* – which gave him all the dry sandy hills, deep canyons and rich river valleys from the Cascade mountains in the east to the

Pacific Ocean in the west, and north to the Oregon border. After his friends, the Americans, won control of California in 1847, Sutter could probably lay claim to more land than any other individual in the Union. Then, with almost theatrical timing, in January 1848 his head carpenter James W. Marshall went to deepen the tail-race to a water-powered sawmill he had put up on the south fork of the American river, and noticed that the silted gravel was flecked with gold. To find untold wealth on his land ought to have been Sutter's final blessing; instead, it was the ultimate curse. When word of the find leaked out, more than forty thousand gold-seekers swarmed into the area, destroying his farms, roads and bridges, and taking without regard to ownership of land the precious metal hidden in his soil – if it was his.

When Sutter took his claim for compensation to court, doubts arose over the real extent of his property. Grants by the Spanish and Mexican governments were respected under the terms of the 1848 Guadalupe–Hidalgo Treaty, if they were confirmed by United States inspection; but the exact limits of the grants to Sutter were in doubt. He spent the rest of his life trying to get compensation for the losses he had suffered, but since the names of the trespassing miners were rarely known either, every effort failed. In 1880 he died, broken in spirit and impoverished, the same fate he once helped inflict on the original occupants of his territory. American measures drove out Mexican as surely as Mexican measures had extinguished the native Californian.

All through the former Mexican territory, similar land grants came under United States inspection. In California, almost a third were overturned. In New Mexico and Arizona, fraudulent claims competed with the genuine and the doubtful. Of eleven million acres claimed under Spanish or Mexican title in Arizona, only 121,000 acres were confirmed – but ten million of those denied were claimed by James Addison Reavis, 'the baron of Arizona', whose frauds eventually secured him a six-year jail term. In New Mexico, Lucien Maxwell used forged land titles to claim 1.7 million acres with much greater success, since he later sold them to British investors for $650,000.

Against this background, the precautions taken by Stephen Austin in Texas stand out for their far-sightedness. The Mexican government wanted to encourage settlement as protection against hostile Apache and Comanche war parties, and it made large grants of land to individual contractors or *empresarios* who were prepared to bring in other families as colonists. Austin was familiar with metes and bounds surveys, and having seen the confusion over land boundaries which 'still exists in Kentucky, Tennessee, and many other states', he made the critical decision to have the Texan land surveyed 'regularly and accurately' under his own direction, rather than 'to let each Settler run his lines as he pleased and mark them or not'. Three different governments were to have control over the land he measured out in careful rectangles. Neither the Texan nor the United States governments ever questioned his or the other American *empresarios*' title to their property. And it was the attempt of the Mexican government in 1835 to challenge that unquestionable right to their land that drove the Texans to fight for their independence.

When Texas was annexed by the United States ten years later, it retained its public land; but because Austin, followed by other *empresarios*, had divided his estates into square or oblong parcels, Texas avoided the litigation-haunted metes and bounds system that dogged other southern states. The advantages inherent in the square-based federal land survey gave the state's economy a vigour its neighbours lacked. After statehood, fifty million acres were sold to support education, thirty million given to railroad companies, three million to the builders of the state capital, and the rest distributed as military bounties, or allocated to squatters and homesteaders. By 1898, Texas had disposed of all its public land, but in the process had transformed much of it into productive capital that helped finance its fledgling oil industry.

In New Mexico, by contrast, the land survey was dogged by corruption and the difficulty of disentangling fraudulent claims from the genuine and often complex rights of local people to land and water that were derived from traditional use or from Spanish and Mexican grants. As late as 1890, the Surveyor-General of the Territory observed, 'certain title to the land is the foundation of all values. Enterprise in this Territory is greatly retarded because that foundation is so often lacking.' Since no real market in land evolved before the twentieth century, the capital to develop the New Mexico economy had to come from outside.

In all the region covered by the Guadalupe–Hidalgo Treaty, it was California which was quickest to establish the essential framework of measurement. On 17 July 1851, United States Deputy Surveyor Leander Ransom struggled to the summit of Mount Diablo, east of Oakland, and dug a hole in what he described as the 'haycock shaped' summit, marking the initial point of the first meridian in California.

'From the top of this mountain a beautiful prospect is opened before you,' he wrote in his report, and went on to describe the sweeping view from the glitter of the Pacific Ocean in the west across the tremendous gap of the Golden Gate round to the Vaca mountains in the north and eastward up the Sacramento river and the nearby valleys. 'These valleys and the ravines and hills surrounding them are mostly covered with a thick set of wild oats growing from 4 inches to as many feet in height,' Ransom noted. 'The wild oats afford abundant pasturage to the extensive droves of cattle and horses that are scattered abroad over this magnificent range, and also to herds of elk, antelope and deer that abound here.' Curling suburbs and square fields have changed that pristine scene, but the shape of the land so intricately indented by bays and estuaries remains, and the rhythm of its rise and fall from cultivated smooth valley to bare, crumpled ridge is still as it appeared to him.

By the 1860s, when the survey based on the Mount Diablo meridian covered much of northern California, the goldfields were largely exhausted and the most productive acres were the drained, fenced farms and orchards of the Sacramento and San Joaquin valleys. Land once given over to cattle had been improved to the point where cereals like rice and wheat could be grown, and in more favoured places there were orchards of plums and apples, or rows of vegetables. The survey gave farmers secure title to their land, but it also allowed government to function efficiently. As early as 1860, California's Surveyor-General was urging the United States surveyors to define the border with Nevada as quickly as possible. 'There are many settlers in the valley along the border who have never paid taxes,' he wrote. 'They are undoubtedly in California. The amount of taxable property is considerable amounting to several millions.' All it needed to make the money California's was a line drawn in the sand.

The link between Leander Ransom's straight line and taxes is invisible, but its effect on farms in California's Great Central Valley cannot be missed. The valley runs roughly north-west to south-east, but the citrus orchards, lettuce fields, avocado groves, vineyards and asparagus beds are all aligned strictly by the cardinal points of the compass, north to south and east to west. The yellow dirt-tracks that form their boundaries appear at regular one-mile intervals following the surveyor's lines in a neat rectangular plaid.

The same fight with geography occurs in Los Angeles, which should be tilted west of north to align itself with the coast and with the Santa Monica mountains. To an outsider the contest is bewildering. Sometimes geography wins, and streets run parallel to the shore or a ridge; sometimes the victor is the survey. Wilshire Boulevard, for example, starts off looking like a geography street as it runs north-west from the city centre, but then a hidden force suddenly pulls it due west, and the whole area on either side swings over, desperately trying to align itself with the invisible meridian and base-line. Over most of Long Beach, the San Fernando valley and central Los Angeles, the meridian wins out, but between Beverly Hills and the ocean, geography comes into its own, and on the fault lines between the two systems, strangers can lose themselves for weeks.

'Surveying in California is a different operation in many respect from what it is in the other States of the Union,' Ransom reported, and explained that gold fever had inflated wages so much that even a humble chainman expected to be paid up to $100 a month, compared to $15 in the prairie states. Nevertheless, in one fundamental aspect, it was and remains undeviatingly the same. A flight from Los Angeles to Kansas City, touching down in Phoenix, Arizona, offers a glimpse of what makes the United States land survey one of the most astonishing man-made constructs on earth. In all the flat land east of Los Angeles, the same struggle can be seen to fit squares of housing into valleys and canyons of every shape but square. The artificial pattern nearly disappears in the desert and dusty red sierras, but high up in the mountains it emerges again in patches of cultivated bottom land where the edges of rectangular fields are aligned with the cardinal points of the compass. All at once, looking down through the clear air, a surveyor's straight line can be imagined, drawn west to east along the base of one of those fields, running invisibly over the bare rocks and dry earth, then coming into sight again blacktopped, graded, rollered and ruled plumb down the centre of a street in Phoenix. And when the aircraft takes off heading east again, the line is still there, stretching ahead as a street, along a rank of shopping malls, faster than a Boeing's shadow, the edge of an industrial park, the limit of an executive housing development, a desert track, straight as a die, suddenly ending in red cliffs and obliteration, until ten or a hundred miles further east, wherever people have settled, it is reborn as a section road or the boundary of a trailer park.

'It was then in a kind of way that I really began to know what the ground looked like,' wrote Gertrude Stein in 1937, taking her first flight over the United States after years in Paris where the straight-edged images of Picasso and Mondrian had educated her eye; 'quarter sections make a picture and going over America like that made any one know why the post-cubist painting was what it was.'

The aircraft eats up in less than three hours the 1300 miles from California to Missouri that the public land survey teams crawled over for more than half a century. The showpiece of their efforts lies in the Great Plains, where the checkerboard of squares permeates the landscape, the economy, the very outlook of those who live there. To surveyors this is the area controlled by the fourth, fifth and sixth principal meridians. Over land that is never quite flat, but rarely rises to bluffs more than a few hundred feet high, and mostly billows gently like a gigantic sheet in the breeze, surveyors like W.J. Neely found the ideal raw material for their art. Whereas poor William Burt sometimes struggled to make forty links in a day through the Upper Peninsula, Neely regularly covered 480 chains, or six miles, a day as he surveyed 'rolling, first-rate prairie, destitute of timber' in South Dakota. Drawn through Minnesota, the Dakotas, Iowa, Nebraska and Kansas, the surveyors' rectangles became the dominant feature of the land.

South from Fargo, North Dakota, the section roads cut each other at exactly one-mile intervals. The ground is so level that a pick-up truck travelling a parallel road a mile away seems to float, red metal above gold stubble on black earth. In his book *Take the High Road,* written in 1939, the pilot Wolfgang Langeweische wrote approvingly that it was 'just what a pilot wants a country to be – graph paper. You can head the airplane down a section line and check your compass. But you hardly need a compass. You simply draw your course on the map and see what angle it makes. Then you cross the sections at the same angle.'

West of the Ohio, most state boundaries are defined either by rivers or by meridians and parallels, but in determining the shape of the Great Plains territories, Congress was so influenced by the land survey that, wherever politics and geography permitted, it made the straight line king. It carved out two states, Wyoming and Colorado, to a perfect box shape, each being four geographical degrees high and seven degrees wide; it awarded three degrees of longitude each to Kansas, Nebraska, and South and North Dakota, then stacked them up like drawers in a filing cabinet between the thirty-seventh parallel and the forty-ninth. Alongside them to the left, it gave four degrees of longitude each to the dryer states of Colorado, Wyoming and Montana, and piled them up from the thirty-seventh to the forty-ninth with the same geometric regularity. Only the Idaho–Montana border betrays a wanton crookedness.

What neither the map nor the view from an aeroplane can reveal is the tension between these artificial shapes and the environment. From the Côteau des Prairies, a long escarpment marking the edge of glaciated plains in South Dakota, there is a view of squared-off prairie – fields, farms, windbreaks, section lines – stretching to the northern and western horizons and all obeying the survey; the sheer expanse of it is as moving and terrifying as an army on parade. Yet on the stillest of autumn days, another power made itself felt. The dry grasses rustled with it perpetually, an insistent pressure of the air that came from far off, no heavier than breathing, but passing with irresistible momentum from the north to the warm south. The very gentleness of it was sobering. Out there a breeze would shake you on your feet, a storm would knock you flat. It carried the heft of a continent.

The violence of prairie weather was something appalling to

settlers. At first it seemed that life could hardly survive the seasonal extremes of temperature, with forty degrees or more of frost in winter, and weeks when the summer heat never dropped below a hundred degrees. When the first arrivals did begin to farm, nothing had prepared them for the spectacular tornadoes and blizzards that burst out of a clear sky and wheeled in giant vortices through the open land. Only after they had been there for some years did they realise that the real threat lay not in these furious eruptions but in the slower, imperceptible pattern of drought and rainfall. According to Lorin Blodget's ground-breaking study *Climatology*, published in 1857, the hundred-degree meridian running through the Dakotas, Nebraska, Kansas and Oklahoma marked the critical boundary – on a rainfall map the land to the west of it was labelled ominously 'the Desert Plains'.

In 1862, the Union Pacific railroad was authorised by Congress to start building westward from 'the hundredth meridian of longitude from Greenwich' near Omaha, Nebraska. It was financed by government bonds, and by grants of public land up to twenty miles deep on alternate sides of the track – the total came to almost thirteen million acres. In that same year of 1862, President Abraham Lincoln fulfilled a campaign promise by signing into law the Homestead Act, which enabled anyone to acquire 160 acres of surveyed land simply by settling there and improving part of it for five years – that is, by building a cabin and ploughing the soil. A $15 filing fee then made the homesteader a property owner. The history of the prairie states was shaped by the conflict between those two events and the climate.

Land speculation had never gone away, despite a disastrous slump in 1837, but the railroads brought it back in spectacular fashion. The boast of William Ogden, President of Union Pacific, was calculated to encourage others to follow his example, but it was also probably true: 'I purchased in 1845 property in Chicago for $15,000 which twenty years thereafter was worth ten millions of dollars. In 1844, I purchased for $8000 what eight years thereafter sold for three millions of dollars and these cases could be extended almost indefinitely.'

What these profits concealed was the fate of earlier owners. In Sugar Creek, 180 miles south-west of Chicago, few of the squatters who had moved onto the Kickapoos' land in the 1820s succeeded in buying their holdings after Angus Langham had drawn his squares. Most packed their wagons and drove away. Yet the settlers who had patented and paid for claims to a forty or a quarter-section hardly did better. At one point, surly, stump-legged Robert Pulliam, the definitive pioneer, owned 560 acres – his original 480 and ten years later another eighty on the other side of the creek, where he planned a dam for a new water-mill. It should have been the classic transition from land to business, but he had paid for it all with borrowed money, and when his lenders called in the loan, his property evaporated.

It reappeared in the hands of people like the shrewd and careful farmer Philemon Stoute Jr, who in 1846 inherited 350 acres from his father, and in 1881 owned 2300 acres in Sugar County. His success came from hiring hands to do extra work and investing his profits in quarter-sections which he rented out to tenant farmers. Most of his hired help and his tenants were supplied by the families of settlers who had lost their farms following the 1837 slump. Jefferson's democratic dream was already beginning to look like Wakefield's hierarchy when the railroad companies and the Homestead Act re-ignited the old ideal.

East of Chicago, the railroad tracks ran between cities, but to the west they were being extended into empty land. They needed people to fill it, to produce cattle and crops for transportation, to push up the value of railroad acres and, with the right sweeteners, to buy them in preference to the government's free 160-acre quarter-sections. 'You can lay track through the Garden of Eden,' James J. Hill, founder of the

Great Northern railroad, pointed out. 'But why bother if the only inhabitants are Adam and Eve?'

As an inducement the companies offered a dream based on the 'forty'. A railroad forty could be had for $100 cash, or $25 down and the balance at 6 per cent interest over three years. It made more sense than homesteading. The land was close to the track. The railroad provided a flat-packed, pre-fabricated cabin. It was easy to ship produce out and bring goods in. But the forty was just a grubstake. Like prudent bank managers, the railroad advertisements urged settlers to start with the minimum and 'build up gradually'. There could hardly be any doubt that they would succeed. 'Why emigrate to Kansas?' ran an advertisement in *Western Trail* published by the Rock Island railroad. 'Because it is the garden spot of the world. Because it will grow anything that any other country will grow, and with less work. Because it rains here more than in any other place, and at just the right time.' The warnings in *Climatology* could be ignored. As the soil was cultivated, it released moisture – 'The rain follows the plough' was the phrase used.

'Four thousand and Four Bushels of Corn from One Hundred Acres', the Chicago and Northwestern railroad company boasted of its land in South Dakota. 'Alfalfa is the best mortgage-lifter ever known,' wrote a Great Northern publicist. 'It is better than a bank account for it never fails or goes into the hands of the receiver. It is weather-proof, for cold does not injure it and heat makes it grow all the better ... For filling a milk can, it is equal to a handy pump. Cattle love it, hogs fatten upon it, and a hungry horse wants nothing else.' In an inspired piece of advertising, the Northern Pacific claimed that not only would a Montana farm make its owner rich, but the state's dry climate was so healthy that not a single case of illness had been recorded there in the previous twelve months – except for indigestion caused by over-eating. The Chicago and Northwestern came back with a poster in big red letters proclaiming the superior fertility of Dakota: '30 Millions of Acres of the Most Productive Grain Lands in the World. You Need a Farm.' Entering into the spirit, the Canadian Pacific shouted louder still about Alberta's rich black soil: 'Our lands should yield you annually 100% of the purchase price; and besides they should increase in value at the rate of 20% a year for at least 40 years.'

At a time when Hollywood was nothing more than arid scrubland, the railroad companies were demonstrating the power of a dream. From all over Europe, people responded to the promise that prairie land, north or south of the border, would not only guarantee independence but prove to be a licence to mint money. In Kansas, Carl B. Schmidt, an agent of the Atcheson, Topeka and Santa Fé railroad, brought over no fewer than sixty thousand German settlers himself, while agents of the rival Kansas Pacific fixed the steamer tickets and railroad journeys of thousands more from Scandinavia, Britain and Russia. But by far the greatest number came from within the United States – the restless, landless, ambitious residents of Illinois, Iowa, Wisconsin and other already occupied states to the east. By the end of the nineteenth century the railroad companies had sold about 120 million acres, comfortably exceeding the eighty million acres of free homestead land that were claimed in the same period.

The reality the settlers encountered was not the rural idyll suggested by the advertisements. In the dry or semi-arid lands beyond the corn belt, it quickly became apparent that no one could live off a railroad forty, or indeed a homesteader's quarter. It was too dry to grow crops unless there was a creek or a borehole for irrigation, and running cattle required vast acreages to provide sufficient grazing – five to eight acres per animal was the minimum. A succession of Acts – the Desert Land Act, the Timber Culture Act and others – allowed larger areas to be homesteaded so long as the settler irrigated the soil or grew trees which were thought to generate moisture themselves. They could not alter the bleak reality that the model of the small farm on which the survey was founded could not work in such conditions.

In 1902 Grace Fairchild from Wisconsin followed her footloose husband Shiloh to a homestead in the dry land of the Dakotas. The economics of the farm depended on being able to put cattle out on the unclaimed open range, and to cut hay from bottom land that belonged to no one except the government. But as more and more homesteaders came in, fencing off the range, there was less free government land available. 'We settlers only had 160 acres in the early days,' Grace wrote in a memoir, 'and that is not enough land to support many critters or make a living raising cash crops. We had to increase our holding or get out.'

Shiloh, whose dream was to breed horses, was as impractical and ambitious as Robert Pulliam back in Illinois, but he had a wife who would have delighted the souls of both Jefferson and Hamilton. The former would have liked the way she started a school for her nine children, read books, increased corn yields with new strains of seed, bought a good bull to improve the herd, and showed sufficient independence of mind to keep her improvident husband farming rather than horse-ranching. The latter would have approved the way Grace earned extra cash keeping pigs and chickens, selling butter, taking in guests – parties of up to sixteen would-be settlers herded by land agents – as well as bottling, barrelling and pickling the produce from her garden, on top of caring for her large family. Like everyone who reads her story, both would have been astonished at her stamina.

With droughts lasting long enough to kill the deep-rooted alfalfa, blizzards coming as late as May, and plagues of grasshoppers which usually arrived 'when the drought had us against the wall', the chance of a wipe-out hung over every Dakota farmer. The harsh truth of homesteading in dry country was that survival did not depend just on stamina or good luck but on others' failure. Abandoned claims offered free pasture and hay to the farmers who remained. When the land was sold to pay back-taxes, anyone with spare cash or credit, like Grace Fairchild, could acquire it cheaply. After forty years, her homestead had become a ranch of 1440 acres, and she had put most of her children through college on the proceeds.

Grace consciously sought out the wider horizons that education offered, but even for her a prairie farm threatened to become a prison. In the early days, she noted, 'most of our neighbors lived 90 miles east of us in Fort Pierre, or twelve miles south of us on the Indian reservation'. Even when the land filled up, her nearest neighbour was a mile off, and her best friend, Mrs Kurzman, who helped deliver several of her babies, had a claim four miles distant. 'In no civilised country have the cultivators of the soil adapted their home life so badly to the conditions of nature as have the people of our great Northwestern prairies,' wrote E.V. Smalley in the *Atlantic Monthly* in 1893. 'Each family must live mainly by itself, and life, shut up in the little wooden farmhouses, cannot well be very cheerful . . . An alarming amount of insanity occurs in the new prairie states among farmers and their wives.'

What Smalley blamed were the squares. Each settler family tended to build near the centre of their holding, so that no field was too far from the house. Because the survey allocated no land for roads, they usually ran along the section lines, which left the average farm a quarter of a mile from the nearest road. On two neighbouring 160-acre claims, that meant half a mile between farms as the crow flew; but by road, down the track to the section line, along the section line and up the neighbour's track, it was closer to a mile. Smalley wanted all the owners of farms in four quarter-sections to 'agree to remove their homes to the center of the tract', and thus create a village with their land redistributed outwards from the centre. Such an arrangement would, he argued, attract people 'of such a sociable, neighborly disposition as would open the way to harmonious living'.

In seventeenth-century Massachusetts, even in eighteenth-century Marietta, that system of in-lots and out-lots radiating out from a town or village centre would have seemed obvious. The pattern had changed while the survey was still releasing the forest land east of the Mississippi. With a square of woodland to clear, most settlers started at the centre. A popular sequence of prints published in the 1850s by Currier & Ives expressed their ambitions. The first picture shows a wooden cabin surrounded by the forest; in the next the cabin has grown to a little cottage standing in a clearing; by the third it is a farmhouse with a yard in front and behind it rectangular fields lined by woodland; and the last portrays a two-storey mansion surrounded by farm buildings, a squared-off garden and straight-edged fields stretching away to the horizon. No other building is in sight.

In Quebec and Ontario, by comparison, the long lots with their narrow end fronting on a shared road encouraged the settlers to build their houses by the road with their farmland stretching out behind. Thus, instead of the typical American farmhouse in the middle of a square section of land with the nearest neighbour a mile or more away, the Canadian farmer usually had seven or eight neighbours within half a mile on either side. Even in the prairie provinces, the Canadian survey had allowed a strip of one and half chains or ninety-nine feet for roads through the townships, and this tended to draw houses together. The United States prairie farmer endured a uniquely solitary existence where progress depended on an individual's ability to impose his or her will upon the land.

The cost was not only human. Square farms carved out of the country regardless of drainage slopes were vulnerable to erosion, and it was from the dry plains, ploughed and planted to exhaustion when prices were high during the First World War, that the soil blew away in the great dust storms of the

1930s that darkened the daytime sky 1500 miles away. That catastrophe represented the most visible sign of the reckless exploitation of the land's resources. Great forests had been burned to clear the soil for crops, the buffalo which had roamed as far east as Georgia in herds ten thousand strong were on the edge of extinction by 1875, and the passenger pigeons that once flew in flocks so great they took hours to pass and darkened the sun had vanished altogether.

When nineteenth-century conservationists like John Muir and John Burroughs began to argue for the social and spiritual benefits of using the land more gently, they faced a particularly American obstacle. While most Western nations had land laws restricting individual property rights in favour of social needs, the United States had the opposite. Combining the legal concept of 'fee simple' with the Fifth Amendment – 'nor shall private property be taken for public use, without just compensation' – the law protected the property owner from almost all government interference so long as taxes were paid. To a degree unknown in the rest of the world, Americans were monarchs of their property, entitled to do almost what they pleased – until it infringed the rights of other property owners. 'There is as yet no ethic dealing with man's relation to land and the animals and plants which grow upon it,' wrote Aldo Leopold in *A Sand County Almanac* in 1949. 'Land, like Odysseus' slave-girls, is still property. The land relation is still strictly economic, entailing privileges but not obligations.'

Yet the sheer beauty of a place like the Yosemite valley made the conservationists' argument for them. When it became the first United States National Park in 1890, it was a symbol that a different way of thinking about the land was emerging. Here the wilderness was more than about-to-be property: it contained 'natural curiosities or wonders', as the park's definition put it, worth keeping for their own sake. In 1906 Teddy Roosevelt made the idea a gigantic reality when he set aside almost two hundred million acres of the remaining public domain for forest and national parks. For ranchers, and lumber and mineral companies, this new category of land was and remains an anomaly. Having been surveyed and measured, it ought to have become property and been used for farming, timber-felling or mineral extraction. That was the Currier & Ives theme. Yet for the first time since Thomas Hutchins crossed the Ohio river, the pattern had been broken, and the tension of that underlying conflict still surrounds the public lands, and the Bureau of Land Management which is responsible for them.

Beneath the last image of a neat and prosperous farmhouse in the Currier & Ives series of prints ran the legend 'The Land is Tamed'. That is what comes to mind as you traverse the Nebraskan prairie, seemingly flattened beneath the immense weight of the sky, its wheatfields smoothly harvested and neatly compartmented by section roads. To the horizon in every direction are the cathedral spires of grain silos, usually in threes, the tallest for wheat or corn, two smaller ones for soybeans and other supplementary crops. It is a picture of regularity and order. The farmhouses and outbuildings squat behind their shelter-belts of cottonwood, each with three or four immense harvesters, drills and cultivators parked nearby. To make full use of the harvesters' power, fields often stretch the section's entire one-mile distance, because at that length the machines can work at over fifty acres an hour, 25 per cent faster than in fields only half a mile long. It should be a scene of rural peace. The land is tamed.

But the sense of stability is illusory. The struggle between crops and weather is fought out every year, and dozens of abandoned farmhouses point to the slim margin between success and failure. The isolation, which telephones and television reduced, has been ratcheted up by the machines that have replaced the human face of farmworkers. The squares continue to foster a sense of independence, as Jefferson hoped, but they also create an unremitting competition in which those with stamina and enterprise are rewarded at the expense of those who lack one or other.

It is the right place to end the epic that caterpillared its way, twenty-two yards by twenty-two yards, with Gunter's chain and Burt's compass, and the foreman calling 'Tally!' and the chainman driving in the peg, and the surveyor scribbling in his notebook, and the flagman and the moundman putting in posts every half-mile, right across the American continent until all of it west of the Ohio was entered on a plat. 'The magnitude of the greatest land-measurement project in history is mind-boggling,' wrote the geographer Hildegard B. Johnson in 1977. 'Never was so much land surveyed in so short a time under the same standardised methods . . . one marvels at the determination with which these men threw and retraced their lines. Still their role as civilian heroes of the frontier is largely ignored in the history of the frontier.'

There were mistakes and frauds. Disputes over the line of the California–Nevada border continued until the 1980s. The redwood area of north-west California was imaginatively surveyed by John Benson and a syndicate of accomplices working in the bars of San Francisco with the help of maps bought from the Coast Survey, and others did the same in Colorado and Utah. Fraudulent land claims plagued the survey in Arizona and New Mexico, holding it up until they were settled. Nevertheless the survey in the forty-eight lower states was virtually complete by the 1930s. Most of Alaska had still to be measured – and parts are still unsurveyed – but the great majority of the surveyors' work was concerned with checking the accuracy of original surveys or replacing wooden posts and other markers that had rotted, been burned or otherwise disappeared.

Since 1785 the land mass of the United States has grown to 2.3 billion acres, and of that total, 1.8 billion acres spread across thirty-two states have been at one time in the public domain. Well over a billion acres have been transferred to individual ownership, with seven million remaining in state and federal government hands. In economic terms alone, it represented the greatest orderly transfer of public resources to the private sector in history.

Somewhere west of the Côteau des Prairies, on a line running down through the dry-land prairie, there is a boundary in that history. This is the open range once occupied by the Sioux and the buffalo, and then by the drover-cattlemen whose semi-feral longhorns grazed its short grass before being rounded up for the stockyards. In the 1870s an estimated eleven million cattle roamed free on what was effectively common land. By then the railroads were bringing in the range's new owners, but lack of timber prevented them putting up rails to mark out their property, though some grew fences of Osage orange trees.

In 1873, Henry Rose, who had a farm sixty miles west of Chicago, invented barbed wire, and immediately it became possible for the range to be enclosed along the lines shown on the surveyors' plats. In 1880, fifty thousand miles of fence were in place and over 200,000 tons of wire were being sold each year. Within another five years sixteen million acres had been fenced in, and the wagons of homesteaders loaded with wire and posts leapfrogged past existing claims out into the short-grass prairies. In 1890, the Superintendent of the census reported that, 'Up to and including 1880 the country had a frontier of settlement, but at present the unsettled area has been so broken into by isolated bodies of settlement that there can hardly be said to be a frontier line.'

This was the remark that prompted Frederick Jackson Turner to study what the frontier had been. His theory about the amorphous frontier might have been flawed, but the census was right – a critically important shift had occurred. As the farmers planted their barbed wire quadrangles across the range, they came into ferocious conflict with the range cowboys. Such a battle could have only one outcome. The entire edifice of the law – and with it every sheriff and US marshal who ever became a white-hatted movie hero – was on the side of property; in other words of those whose land was measured and entered on a plat.

The cowboys were the last in a line of unpropertied people to claim the land – and to be defeated by the law. By the 1890s they were on their way to join the Mexican *pueblo* farmers, the Native Americans, the south African Xhosas, the Scottish Highlanders, the English Levellers and Elizabeth I's sturdy rogues and vagabonds among the dispossessed. The long march of enclosures that had begun in England in the 1530s had reached its culmination 350 years later in the dry-land prairies of America.

EIGHTEEN

Four Against Ten

FOUR HAD BEEN THE KEY to it all. That was the number that linked the sides of a township to the old medieval units of the rod and the acre. It was a number the mind could immediately recognise, and a quantity, as Jefferson understood when insisting on square containers, that could be easily visualised and simply tested. To the land-dwellers who relied on them, the four-based measures felt organic, almost instinctive, a sound, practical, hands-on, symmetrical way of measuring; and with it went a way of thinking that valued those qualities. Out west the idea of being four-square – 'a four-square guy', 'a square deal' – came to epitomise everything that was desirably solid and reliable, and by the same way of thinking a square represented all that the unpropertied, sinuous world of jazz despised.

'Foure graines of barley make a finger,' went the rule that Elizabethan England followed, 'foure fingers a hande; foure handes a foote.' And by a freak of history, those ancient quantities had left their mark on the character of the most modern society on earth. Edmund Gunter's chain, the first United States standard of measurement, carried the old English system of land measuring, the eight-furlong mile, the 4840-square-yard acre and the four-based way of thinking, into every corner of the land, and planted it deeply in the American psyche.

Confronted by its spread, Ferdinand Hassler had standard-ised the other weights and measures that went with it: the sixteen-ounce pound, the eight-pint gallon and the four-peck bushel. As each square township out on the frontier filled with people and enough squares within a territory petitioned for statehood, an exact, precision-machined replica of the federal government's four-square weights and measures was sent to the new state capital.

Among the huge variety of state laws and constitutions, this national system of weights and measures created one single market for American industry's goods. It permitted manufac-turers to standardise and mass-produce everything from agri-cultural machinery to small arms, typewriters and sewing machines. For those who wished to acquire wealth, industry became the new frontier. In just twenty years, between 1879 and 1899, the value of industrial products almost tripled to $13 billion a year. The fortunes acquired by Andrew Carnegie from steel and John D. Rockefeller from oil proved beyond doubt that the day had passed when land was the prime source of productive wealth.

As it had been at the very start of the industrial age, the ability to measure precisely was a source of power. In the 1760s James Watt had struggled to get tolerances of a hundredth of an inch, but before the end of the century Henry Maudsley, the British designer of the screw-cutting engine lathe, refined the benchmark to thousandths of an inch, and in the 1850s Joseph Whitworth's routine production was measurable in ten-thousandths of an inch. Metal, power-driven lathes working at this tolerance created the power-looms and locomotives that drove forward the second stage of the Industrial Revolution. By the 1880s Pratt & Whitney were supplying a market for commercial machines capable of checking precision gauges to one hundred thousandth of an inch.

That degree of accuracy was good enough for the metal-bashing end of industry, but the late-nineteenth-century technologies of electricity, telegraphy, chemicals and pharmaceuticals demanded even higher standards. To be used effectively, their quantities had to be assessed with extraordinary fineness.

'Although the electrical is the latest developed branch of engineering, it is the most exact in its measurements,' a technical journal boasted in 1903, 'currents from as small as 1/10,000,000 ampère and less, up to 2,000 or 3,000 ampères being easily and accurately measured ... in electrical work, [these are] powers large enough to operate a train of some hundreds of tons weight, or powers so small that no ordinary electrical device could measure them.' The precision this implied was breathtaking, but it was not just the exactness of the measurement that would have shaken a chain-wielding surveyor to the steel-capped toes of his boots. The unit used, the ampère or amp, contained a secret – it was metric.

That the metric system had survived at all was astonishing. For the first thirty years of its existence, it had seemed perpetually on the point of being abandoned. The revolt against it had begun in France almost as soon as the law of 18 Germinal Year III, or 7 April 1795, had made it the only legal measure that could be used. 'The centuries-old dream of the masses of only one, just measure has come true!' the Committee of Public Instruction proudly announced. 'The Revolution has given the people the metre.' Whether they wanted it was another matter.

Despite the publication of numerous tables giving equivalents between the old and the metric measures, the country was utterly unprepared for the new system. Customers could not understand the metric units, and traders immediately took advantage of the change to increase profit margins. In the great agricultural market of Les Halles in Paris, grain merchants measured sacks of oats by the small bushel but charged them at the price of a large hectolitre. The same trick was played in bakery and grocery shops, where two sets of scales and measures were used, and goods would be sold in old measures under metric names, or vice versa – whichever paid best.

Where there was no fraud, there was confusion, particularly when it came to decimals. In the fashion shops, where lengths of cotton and linen had been sold by the *aune* or ell, it ought not to have been difficult to substitute the metre, which was only a little longer; but the system contained a snakepit of lesser numbers. Everyone knew what half an *aune* or a quarter of an *aune* looked like, and they could halve or quarter the metre in the same way. Given time they could even estimate how long half a quarter of a metre should be, but no one, not even the ink-stained book-keeper upstairs, knew that it was the same as 0.125 of a metre, or, still more bafflingly, 12.5 centimetres.

An alarmed government circular blamed the unpopularity of the system on criminals. 'The general public suffers from dishonest practices,' it claimed, 'and so loses faith in a system that seeks above all to benefit it.' But as the *cahiers de doléances* had made clear, what the public wanted was the old system made uniform, one *toise*, one *aune*, one *quintal* throughout the nation. Instead they had been given what Witold Kula, the Polish science historian, called 'a strange new-fangled measure allegedly relating to the very land people walked but in a manner that no one understood'. The Greek and Latin names, all the kilos and centis that the scientists loved, simply sounded so foreign that French patriots refused to say them, asking instead for the old *livres* and *aunes*.

But the foreignness of the metric system went deeper than names. It took uniformity to a degree that no lay person could immediately comprehend. The traditional measures had variety because they related to different activities. Cloth was measured by the ell or the *aune* because it was natural to hold it and stretch out the arm to full length. A journey was measured by the yard or the *toise* because the road was walked.

Land was measured by the acre or the *arpent* because that represented work. Now people had to separate the measure from the activity altogether and deal with an abstract unit that, as Kula observed, 'would be equally applicable to textiles, wooden planks, field strips and even to the road to Paris'. What underlay the popular dislike of the metric system was a very modern anxiety, the sense of alienation from the natural world.

In 1799 the system received endorsement from a French-led international commission made up of subservient states like Switzerland (represented by Hassler's friend Johan Tralles), Batavia and Lombardy. Following their approval, per-manent standards for the metre and kilogram were made from Joseph Dombey's platinum based on the measurements of the meridian by Delambre and Mechain. These standards were deposited in the Archives of the Republic, and copies made for members of the commission. When Napoleon seized power and crowned himself Emperor in 1804, he threw all his power and prestige behind the metre. An edict issued in the following year announced: 'It is most definitely his unalterable wish to maintain the new system of weights and measures in its entirety and to accelerate its extension throughout the Empire.' But neither international prestige nor imperial decrees could per-suade the French to love the metre.

The Emperor did not have much time for decimals himself. In his army the official daily ration for cavalry horses remained a quarter of a *boisseau*, the artillery measured the calibre of their guns in *pouces* and *lignes*, and when his military engineers crossed the Beresina river into Russia, they estimated its width at forty *toises*. By then, Napoleon had begun to retreat in the face of France's massive sullen resistance. In February 1812 his much-feared Minister of the Interior, Jean-Pierre de Mon-talivet, informed the prefects in charge of France's eighty-three departments that the Emperor had come to recognise that 'metric units are not suited to practical daily needs. The exclusive employment of the decimal system may suit book-keepers but is by no means well adapted to the daily dealings of the common people who have much difficulty in under-standing and applying decimal divisions.'

The solution was to introduce what was called the *système usuelle*, combining metric and organic units. The *toise* came back, but now measuring two metres, as well as six *pieds*. The *livre* reappeared, representing both five hundred grams and sixteen ounces, while the new *pied* could be thirty-three centi-metres or twelve *pouces*.

When Napoleon was overthrown and the Bourbon mon-archy restored in the shape of Louis XVIII, the metric system, brought in by the hated Jacobins who had guillotined the last King of France, might well have been abolished. It had almost no friends, yet somehow it survived. The explanation lies in a snobbish aside made by the vicomte de Chateaubriand in his *Mémoires*: 'Whenever you meet a fellow who instead of talking of *arpents*, *toises* and *pieds*, refers to *hectares*, *mètres* and *centimètres*, rest assured, the man is a prefect.' And in case any of his readers had forgotten what a prefect was, the vicomte help-fully defined him as 'a person characterised by petty tyranny, a bureaucrat concerned with conscription' – and, it should be pointed out, with responsibility for weights and measures.

Among the briefing papers put in front of Louis XVIII after his coronation in 1815 was one from his new Minister of the Interior, to whom all prefects reported. 'Sire!' it began, 'The uniformity of weights and measures has long been desired in France, and your royal predecessors sought to establish it. I presume therefore that your Royal Highness will wish to uphold an institution that accords so excellently with his great ideas of public utility and whose effective spread is in any case by now very much advanced.'

Whatever the new King's private opinion, he was a realist. He might claim to be an absolute monarch, but he ruled through the vast, centralised government machine that had grown out of the Revolution and the Napoleonic Code. It was the machine that wanted the metric system. As Napoleon himself had been forced to admit, the simplicity of calculating in decimals suited book-keepers – and prefects, and bureaucrats, and government officials of every kind. Swallowing his reservations, Louis agreed to keep the metric system alongside the customary measures. 'The important lesson France has taught the world,' remarked a cynical friend of Kula, 'is the effectiveness of centralised administration.'

It was not wholly a coincidence that the unpopular metric system spread across Europe in concert with the growth of government bureaucracy. Each of the client states which first received the metric system – Lombardy, Switzerland and Batavia, which was to become part of Belgium – reacted in the same way as France, accepting it officially and as a symbol of modernism and democracy, while rejecting it at almost every other level. But like France, they too discovered that the system was a one-way street. Introducing it created chaos; trying to go back to the customary measures produced still more. The indecision lasted for a couple of generations, and it was not until the 1850s that their governments found the courage again to make the metre the only official measure.

It was Prussia that really learned the French lesson about centralised administration and thus set the model for metrication. Napoleon had lost faith in the system by the time he crossed the Rhine in 1806, but in 1868 the autocratic Prussian Prime Minister, Otto von Bismarck, set about modernising the most traditionally hidebound state in Germany. In what was called 'a revolution from above', he reformed its administration, its public education and its social security – and replaced the old *Zoll* or inch, *Pfund* or pound, *Morgen* or half-acre, with the centimetre, kilogram and hectare.

Once Prussia, the first truly modern state, succeeded in unifying Germany beneath its banner, the metre and modernism soon spread into Austria and Hungary. Beyond German borders a freshly united Italy caught the spirit and embraced the metric system as a symbol of its new-made nationalism. Scandinavia had adopted it earlier in 1863, and Spain, which had introduced it for the second time without much enthusiasm, now extended it to Cuba and Spanish America, where it took hold among newly independent countries enthusiastically signalling their progressive credentials. In 1870 the delegates from twenty-four different nations met in Paris to agree on international standards for the metre, and even the French, who in 1840 had had it re-imposed as their country's only legal system, became – rather doubtfully – a little proud of it.

Thus, against all the odds, the deeply unpopular, undemocratic units based on Borda's triangulation had survived and prospered. But the system had other allies than bureaucrats. While the chain had been spreading traditional measurements through the American west, the physicists mapping out the vacant properties of heat, light and electricity had decided to measure them metrically. They needed simplicity and exactness, and in their work, pounds and feet were Mexican units, too impractical to be easily used.

The fault lay in the dimensions of the world that science uncovered in the century after 1785. It grew until it encompassed everything from the unimaginably small to the inconceivably large, from molecules to galaxies. To describe these new-found extremes created a need for precision beyond anything that had gone before. In the United States, Hassler found little support for his efforts to push the standards of accuracy to extraordinary limits, but in Europe the École Polytechnique in Paris, and universities like those at Berlin, Glasgow, Turin, Uppsala and Cambridge, as well as scientific bodies like the Royal Society and the Académie des Sciences, created an environment in which precise measurement was recognised as fundamental to scientific experimentation and discovery.

Its importance prompted the great Glasgow physicist William Thomson, later Lord Kelvin, to a famous passage in a lecture in 1891: 'When you can measure what you are speaking about, and express it in numbers, you know something about it; but when you cannot measure it, when you cannot express it in numbers, your knowledge is of a meagre and unsatisfactory kind: it may be the beginning of knowledge, but you have scarcely in your thoughts, advanced the state of science.' (The impact is best caught when read in Kelvin's flinty Scots accent.) Thus, just as the surveyors had once used measurement to create property from the wilderness, so in the nineteenth century the physicists used it to create science from the natural world.

One fundamental need had first to be met. To describe how electrical energy is turned into heat, light, magnetic energy or mechanical power required a new system of uniform measurement. The problem was considered in 1861 by a committee of the British Association for the Advancement of Science, including Kelvin himself as well as two other giants of nineteenth-century physics, James Clerk Maxwell and James Prescott Joule. Any measurement of electrical energy had to be a synthesis of its constituent elements of charge, current, voltage and resistance. The committee decided that each element should be measured separately, using the fundamental principles of length, mass (as weight was now more accurately described) and time. In what was to be a critical intellectual decision, the members of the British committee chose as the units of length and mass not the inches and ounces they used in everyday life, but the metre and gram, later amended to the centimetre and gram. Although time continued to be measured in old-fashioned seconds still reckoned as one 86,400th part of a day, electrical measurement became metric.

The choice was a matter of convenience. A common system of measurement was needed because physicists from at least four different countries had helped uncover the properties of electricity and magnetism. And the ease of converting small quantities to large simply by moving a dot made decimals ideal for measuring the great range of quantities found in the use of electrical power.

The decision of the British Association for the Advancement of Science crowned the metric system as the favourite child of science. Other areas of physics concerned with light, heat and radiation needed measurement, and all were linked by James Clerk Maxwell's equations on electromagnetic waves. A series of international scientific congresses hammered out agreement on the nature and names of the new units during the late nineteenth and early twentieth centuries. As well as distance, mass and time, four other basic elements were recognised – temperature, substance, light intensity and electric current – for which the units were named respectively the kelvin, mole, candela and ampère. From these seven elements, composite units were derived which could be applied to every other measurable force – power, radiation, electric charge, magnetic strength and others – many of them named for the scientist or mathematician who first described what was being measured. Units like the watt and the volt, ampère and ohm, curie and hertz, siemens and becquerel, were the equivalent of physics' hall of fame. And they were all metric.

It was the decimal division and the integration of length and mass that proved to be the system's strength. Its origin, as one ten-millionth of the length of a quadrant meridian, had become unimportant. The true distance from equator to pole, now reckoned to be 10,001,965.7 metres, was almost two kilometres longer than first estimated, and over 1200 metres beyond Delambre's best computation. Even in the nineteenth century, the 1799 bar of platinum in the Archives of the French Republic, from which all metric lengths were derived, was understood to be as much as 1/150th of an inch shorter than it should be. Nevertheless, where the physicists led, others followed, and for similar reasons.

In 1867, surveyors and mappers from around the world

met at a conference in Berlin of the International Geodetic
Association and recommended that a new international metre
based on the 1799 metre be adopted as the standard for all
their work. Governments could ignore science, but maps,
which involved frontiers and national territory, were a different matter. A preliminary series of meetings in 1870 was inter-
rupted by war, but the representatives of thirty nations met
again in Paris in 1872, examined the metre and kilogram
made from Dombey's platinum and kept in the Archives, and
decided they should be the basis for a new set of internationally accepted standards. Three years later, eighteen
countries signed the Treaty of the Metre, which brought into
being the International Office of Weights and Measures that
still supervises metric standards, and created for it an international enclave, a kind of Vatican of the metre, at Sèvres
outside Paris.

The new international standard was introduced in 1889 in
the form of a bar cast in a platinum-iridium alloy, but as close
in length to the 1799 metre as possible. Instead of being
sawn to a length of one metre, with the consequent risk of
contracting to a shorter length, it was 102 centimetres long,
with two scratches etched into it at exactly one metre's distance. Since it had an error margin of as much as one twelve-
thousandth of an inch, it was outdated almost as soon as it
was made, and before the end of the century physicists were
unofficially defining the length in the more precise terms of
wavebands of light. As the world continued to expand relent-
lessly into sub-atomic quantum particles and stellar regions
beyond the Milky Way, the need for a still more exact definition grew urgent.

It came in 1960, with a new specification based on the
wavelength of light. Careful measurement of the 1889 metre,
itself based on the 1799 bar, showed its length to be equivalent
to 1,650,763.73 wavelengths of radiation emitted by the
orange krypton-86 atom. This became the new definition of

a metre. The accuracy was sub-atomic, a margin of error much less than a millionth of an inch; but for the first time there was no physical standard against which a length could be checked, merely a scientific experiment.

The change was critical. Once released from physical constraints, the metre has become defined with a growing fineness that now far exceeds the imagination of the non-scientist. It is presently equal to the length of the path travelled by light in a vacuum during the time interval of 1/299,792,458th of a second, where a second is equivalent to 9,129,631,770 oscillations of the 133 Cesium atom. Which is not quite as straightforward as a yard once was, but has some advantage in exactness. Using an iodine-stabilised, helium-neon laser built according to international guidelines, the metre's accuracy can be checked with an error margin of plus or minus 0.000,000,000,000,02 millimetres. Ever more precise and less comprehensible definitions of its length will certainly continue to evolve.

Only the kilogram remains a physical reality, an iridium-platinum cylinder made, like the metre, in 1889 and kept in a locked vault in the Sèvres headquarters of the International Office of Weights and Measures. It usually remains out of light and sight, protected against erosion and deterioration by three airtight bell jars. Although it looks remarkably like the copper example that Joseph Dombey carried with him on *The Soon*, it is no longer a unit of weight but of mass.

For most practical purposes the two are virtually the same, but as scientists are irritatingly quick to point out, if you were on the moon your mass would not alter, but your weight would be only one-sixth of the amount on earth. This is because weight measures the force of gravity on mass. Even on earth, a person increases in weight at the pole, where gravity is stronger, and floats more lightly at the equator, where it is weaker. Thus, by the sort of logic that can blow a fuse in the non-scientific mind, the weight of the kilogram is actually determined in newtons, a unit of force (one kilogram is roughly 9.8 newtons), while its mass is simply 'the mass of a particular cylinder ... which ... is preserved in the care of the International Bureau of Weights and Measures in a vault at Sèvres, France'.

Eventually this too will be consigned to history, to be replaced by a definition related to the electrical force needed to lift a kilogram, or the number of atoms in a silicon crystal of that mass. The substitute is needed because, for all its protection, the mass of the Sèvres kilogram is getting larger as a film of atmospheric particles coats it, adding what is effectively the mass of a grain of pepper every decade.

At the eleventh General Conference on Weights and Measures held in 1960, all the metric measures, old and new, were brought together into what is called, in English, the International System of Weights and Measures. Its official title, in deference to the nationality of Jean-Charles Borda, with whom it originated, is *Le Système Internationale de Poids et Mesures*, or SI for short.

That year marked the point at which the globalisation of measures became a reality. The government of every industrialised country in the world signed up to the Treaty of the Metre. All their weights and measures were defined in relation to the metre and the kilogram. Most had gone further and adopted the SI as their official system.

In 1785 James Madison had written prophetically that a scientific, decimalised system 'might lead to universal standards in these matters among nations. Next to the inconvenience of speaking different languages, is that of using different and arbitrary weights and measures.' Much of that inconvenience was now being removed, and the massive expansion in international trade following the Second World War was undoubtedly accelerated by the single language of measures that by 1960 had come to be spoken around the world. In two major areas, however, it still had to be translated into a local *patois*. In those countries that Winston Churchill dubbed 'the English-speaking democracies' and General de Gaulle simply called 'the Anglo-Saxons', the old traditional measures were still firmly in place.

NINETEEN

◆◈◆

Metric Triumphant

ETWEEN AUSTRALIA, Canada, New Zealand, South Africa, the United Kingdom and the United States, there was indeed a bond which arose from language and form of government; but they shared a third and perhaps more fundamental tie. One of the profoundest influences on their legal and democratic systems was that concept of private landed property whose physical reality had been carved out with Gunter's chain. To protect what they claimed as theirs against over-greedy rulers, land owners had struggled to establish a system of individual rights and a form of government in which their interests were represented. Thus for all six nations, property and democracy came together at the same point, the enclosures in Tudor England, and they seemed inseparable from the system of organic, four-based weights and measures on which ownership and exchange depended.

By contrast, the metre had emerged from the same intellectual ferment that enunciated the rights of man as an idea, something innately human and existing independently of property or any other tangible quality. Around the world the metric system had always been introduced from the top rather than in response to popular demand. Each country had its own motives – the Soviet Union went metric in 1918 as part of a programme of cutting ties with its old imperial Russian past; in 1947 a newly independent India announced its intention to switch in order to underline its break from Britain; Japan changed in 1958 in accordance with an overall plan to expand its foreign trade – but there was always one constant: the existence of strong central government with a clear desire for modernity.

As colonial powers, nineteenth-century France and Belgium made the metric system part of what they deemed their civilising role in African possessions such as Algeria and the Congo, while former British colonies like Ghana, Nigeria and Kenya adopted it in the twentieth century as a symbol of their newfound independence. In 1872 the liberal Emperor Pedro II introduced it as part of his programme to modernise Brazil's economy. Mexico had the system imposed three times in a decade – twice by Benito Juárez and in between by the French puppet, Emperor Maximilien – each time in the name of progress. When Mao Zedong and the Communist Party decreed in 1959 that the People's Republic of China would go metric at the start of their ambitious but disastrous modernising plan, the Great Leap Forward, the system circled the globe.

For the English-speaking hold-outs, the question of adopting the metric system consequently carried a wider significance than the simple exchange of one set of measures for another, and for a century every attempt by governments to move towards introduction of the metre had been frustrated.

Despite its resistance to Jefferson's decimals, the United States ought to have been one of the first metric countries. As early as 1866, just nine years after establishing the American customary weights and measures as the federal system, Congress agreed that metric measurements could legally be used alongside it. Six years later the United States became one of the original signatories to the Treaty of the Metre, and an increasingly influential scientific and technological community was pushing the government towards adopting the system. In 1893, Thomas C. Mendenhall, the politically powerful boss of the Coast and Geodetic Survey and thus, following

Hassler's precedent, responsible for weights and measures, persuaded the Treasury that the yard should no longer be defined by the 1857 standard, but in relation to the decimally divided, light-wave determined, scientifically based metre. At that point, the intellectual battle should have been over – the American customary system no longer had any independent basis for its standards.

When politicians appeared more inclined to listen to the argument that a change would confuse customers, particularly the poor and the less educated, the scientists voiced their opinions with growing impatience. 'If the introduction of the metric system is to be accomplished in America,' W. Le Conte Stevens wrote sharply to the *Scientist* magazine in 1894, 'we must act in the light of experience already acquired in Europe which is far more valuable than any amount of theorizing about the effect upon poorer classes who have not yet tried it.'

To galvanise the lawmakers further, scientists pointed across the Atlantic to the astonishing sight of the nation's largest trading partner, Britain, suddenly moving rather faster towards the metric goal. In 1895 a House of Commons committee recommended forcefully that Britain should take the plunge, and was supported by some of the most powerful business organisations in the country. 'A decided advantage will accrue to whichever of the two great English-speaking nations shall first put itself in line with the rest of the world in this, one of the greatest economic reforms of the nineteenth century,' warned another of the *Scientist*'s correspondents that year. 'Up to the present time, we have been on the whole in advance of England.'

The turnaround in the British government's attitude was almost incomprehensible given the country's industrial power and historical distrust of any French innovation outside the bedroom. There was so little support for the metric system beyond an educated elite that an opinion poll conducted in the 1860s concluded that its introduction could only be achieved in countries with an authoritarian regime. When the Treaty of the Metre was presented for signature in 1875, the British government sharply informed its delegate that it 'would not recommend any expenditure connected with the metric system'.

Yet ten years later they had signed the treaty, and in 1897 legislated to allow the system to be used alongside Imperial measures. The 1895 Commons committee had recommended this as the first step to metrication. The second would be to make the metric system compulsory. Backing this was a powerful pressure group made up of business in the form of chambers of commerce, education groups, engineering associations, and exporters. Their motives were various: the teachers gave evidence to the committee that 'no less than one year's school time would be saved if the metrical system were taught in place of that now in use'; manufacturers wanted to be rid of the handicap of having to cope with two sets of measurements in constructing machinery; and business found that trade suffered from the difficulty of supplying metric-sized components for customers abroad.

What had changed everything was the realisation that the long era of industrial dominance built on Watt's steam engine and a host of subsequent innovations was drawing to an end. German steel and chemical production had overtaken Britain's, and the education system imposed there by Bismarck was seen to be superior. A mood of uncertainty gave rise to a conviction in the British government that the country needed to modernise. There was no indication that public opposition had been modified even slightly.

Then in 1899 war in South Africa broke out, a new administration chose different priorities, and the two world conflicts in the first half of the twentieth century distracted government attention from a declining economic base. It was in the aftermath of victory in 1945 that Churchill coined his phrase 'the English-speaking democracies'. He might equally well have named them the acres and ounces democracies. Having rescued the world from the threat of totalitarian dictatorship, their populations saw less need than ever to succumb to metrication.

In 1959 the governments of the six agreed to harmonise their measures in relation to the metric system (a pound was declared to be equivalent to 0.453,592,37 kilograms, and one inch to 2.54 centimetres; the impossibility of making an exact equivalence led to a three-millimetre difference between the international mile, derived from a metric base, and the statute mile of Queen Elizabeth I, which in the United States is derived from the foot used in the public land survey). But while the change bound their systems closer together, it exposed a fundamental weakness.

In Britain and the Commonwealth nations, the ultimate standards of their weights and measures were the metal bar and cylinder constructed in 1845 to replace those melted when the Houses of Parliament burned down. By the 1950s, the latest scientific methods revealed that not only were there minute inaccuracies in their construction, but exposure to the atmosphere was causing a steady, microscopic deterioration in their condition. Either new standards would have to be made or a new definition established. In 1963, Britain bowed to the ravages of time, and legislated to define its yard in relation to the metre and the indestructible, invariable wavelength of orange-red krypton 86. When Australia, Canada, South Africa and New Zealand followed suit, every major country in the world shared the same scientific definition of its weights and measures.

Thereafter the chasm grew wider, as the governments of each of the six hold-outs, urged on by their scientific, business and industrial communities, began to adjust for formal entry into the metric system, while the majority of their populations opposed it.

In 1965 the British government announced a ten-year plan

for conversion, and until 1979 a Metrication Board announced the different areas of business, industry and the administrative structure that had gone metric – glass-makers, schools, government contracts, filling stations, the London Metal Exchange. The alcohol in whisky was expressed as a percentage rather than as proof; the gill in which it was measured followed the pottle and the mutchkin into history. Even Gunter's chain, the very measure that made private property possible, faded from view, accompanied by a bodyguard of rods, poles and perches.

The conversion was voluntary, but in 1973 Britain joined the European Common Market, whose rules were designed to make the metric system compulsory. The period of conversion was elastic, in some cases lasting over twenty years, but sooner or later the change would have to be made. By the time it was wound up in 1979, the Metrication Board reported that except for retail food stores, home improvement sales and road signs, metric measures were either replacing traditional units or being used alongside them. Almost as an aside, its final report noted that just 31 per cent of the population supported the change.

A similar sequence of events occurred in Australia, Canada, New Zealand and South Africa. The process was voluntary up to the point where customers could make their wishes felt. Thus pre-packaged food like breakfast cereals and frozen peas was sold in metric units early on, but fresh food like potatoes and fruit continued to be weighed out in pounds and ounces. Wholesale quantities in the timber trade were specified metrically, but individual planks were ordered and sold in feet and inches. To clear up the confusion, retail trade associations and chambers of commerce began to urge governments to move into this area too, and to pass legislation making the old units illegal. Australia, New Zealand and South Africa had largely completed the process by 1980, and Canada was judged to be 60 per cent complete.

At first the United States marched in step, beginning with a report to Congress in 1971 optimistically entitled *A Metric America, a Decision Whose Time has Come*, which called for a programme to make the country predominantly metric within ten years. Acting with a speed unparalleled in earlier attempts to reform weights and measures, Congress passed the 1975 Metric Conversion Act, setting up a board to 'co-ordinate and plan the increasing use of the metric system in the United States'. In an interesting example of political ambivalence, the Reagan administration closed down the board but approved the 1988 Omnibus Trade and Competitiveness Act designating the metric system as the 'preferred system of weights and measures for United States trade and commerce'. For good measure the Act also required federal agencies to use the metric system in procurement and other business by a 'date certain and, to the extent economically feasible, by the end of fiscal year 1992'.

Under the Act's impetus, large sections of industry which depended on government contracts went metric, and defence procurement in particular abandoned traditional measures. At the other end of the price scale, soft drinks were sold in two-litre bottles, and spirits that had once appeared in fifths and quarts came in 750-millilitre and litre bottles. Medicines were measured in grams rather than grains. In the year 2000, the New York Stock Exchange began replacing the old eighths and sixteenths fractions of a dollar in which stocks were priced, and – logically enough – made use of Jefferson's decimals to mark their price in dollars and cents. New sectors like the computer industry quoted quantities of information bytes in metric notation, going from kilo through mega and giga, eventually reaching exa, zetta and the so far ultimate yotta, or 1,000,000,000,000,000,000,000,000 bytes.

According to the writer Robert Sabbag, one part of the economy had already gone decisively metric a decade earlier. 'Whatever Congress decides,' he wrote in his account of the drugs trade, *Snowblind*, 'the truth is this: the United States of America effectively converted to the metric system in or around 1965 – by 1970 there was not a college sophomore worth his government grant who did not know how much a gram of hash weighed . . . off the top of his head he could go from grams to ounces, and he could tell you how many ounces he got to the kilo . . . And so today, everyone over the age of twelve *knows* there are 28.3 grams to the ounce, and 35.2 ounces or 2.2 pounds to the kilogram.'

Throughout the English-speaking democracies the momentum for change seemed irresistible. Running two systems of measurements was inconvenient for government, expensive for industry, liable to create error in science and advanced technology. All of these fields had an interest in taking the last step in metric conversion and eliminating traditional measures; and since those measures were now based on metric measurement, there was no logical reason for keeping them.

Yet even in countries which had adopted the metric system for more than a century, where children grew up educated in it and thinking in terms of a single unit of length, weight and capacity, a residue of organic measures remained stubbornly in existence. In German markets they sold, and still sell, meat by the *Pfund*, or pound, meaning five hundred grams or 17.6 ounces. In French markets the same unit appears again, this time as *la livre*, while the Danes order in *punds*, the Dutch in *ponds*, and the Swiss in *Pfunds* or *livres* depending on which canton they live in. Heavier quantities such as coal or lumber have kept the hundredweight or *Zentner* (one-twentieth of a ton) alive in Germany, where it is used in place of fifty kilograms, and one metric tonne becomes *zwanzig Zentners*.

The inch, whose death sentence was signed in Scandinavia in 1863, is resurrected every day by plumbers who routinely specify the diameter of pipes and faucets in *tumme* (literally thumbs) if they are Swedes, or *tomme* if Norwegian. Even in Europe's industrialised heartland, Germany, the thumb, or the *Zoll*, is what craftsmen use for small measures. Informally a Swedish farmer will measure his land not by the hectare but by the *tunnland* (roughly 1.2 acres), his Austrian counterpart by the *Joch* (1.4 acres), a German by the *Morgen* (0.6 acres) and an Italian by the *acro* (one acre).

The feature common to all these measures is that they fit everyday activity better than the equivalent metric units, and in personal, hands-on transactions, customers prefer the most suitable, whole-number units. But their survival is due to more than convenience. People insist on using them in direct exchange because it is an unconscious sign of belonging to the community, of knowing the local dialect, as opposed to a stranger who would use the metric term. Here the metric system's strength, its invariable, universal uniformity, becomes its weakness. Like all forms of globalisation, it saps the sense of local autonomy.

The resistance among the six hold-out nations naturally went further, and in the 1980s it forced the governments of Britain, Canada and the United States to abandon plans to impose the system compulsorily. In states like Iowa, Alabama, Missouri and Illinois, where speed limits had been designated in kilometres in the 1980s, the signs were removed in the 1990s. State building codes went from metric back to feet and inches.

In Canada, where metrication had been promoted aggressively with massive publicity in the form of pamphlets, leaflets and films, and a rolling programme covering industry, government and even retail stores, the public reaction halted the campaign in its tracks in 1983, leaving a remarkable confusion of units. Specifications for construction work on government plans were metric, but for private buildings in feet and inches. Weather temperatures were referred to in Celsius, oven temperatures in Fahrenheit, snow in centimetres, wind in miles per hour. The *Washington Post* reported in 2000 that in the country's largest supermarket chain, 'the produce clerks speak in pounds; the fish counter measures by grams; and the meat department maintains a studied bilingualism – frozen turkeys in metric, fresh in imperial. Loose mushrooms are priced in pounds, but the scale available to weigh them measures only in grams'.

A similar confusion reigned in Britain, with television weather forecasts given in metric and Fahrenheit, and doctors recording babies' weights in kilograms but telling their mothers in pounds and ounces. Road signs remained in miles per hour, but petrol was sold in litres. In 2000, when Steven Thoburn, a market trader in Sunderland, refused to quote prices for bananas in kilograms, the city authority took him to court; but elsewhere other would-be 'metric martyrs' were allowed to ignore European Union rules, and a reliable estimate suggested that altogether about thirty thousand shopkeepers had neither replaced nor converted scales that weighed only in ounces and pounds.

'Their slipping back into the old ways served some hunger of the soul,' a commentator in the Australian magazine *Quadrant* suggested of that country's metric refusers. 'It might also have been a healthy instinctive resistance to the bossiness of all revolutionaries; that rigid insistence on conformity with the new dispensation.'

The refusers found an unexpected ally in the new American-dominated computer industry, whose hardware, such as disks, monitors and printers, was measured in traditional units. Software for printers and for complex CAD (computer-aided design) applications also used inches and fractions of inches, which created unexpected problems in metric mode. American architects, for example, who use tolerances of one-quarter, one-eighth or one-sixteenth of an inch in their work, found the software for technical drawing ideal for their requirements, but metric architects complained that the software insisted on approximating their more precise millimetres to the nearest fractional equivalent, and so threw out their calculations. Even the computer's basic information code, based on binary arithmetic and thus having to be counted in twos, defies decimalisation, so that there are 1024 bytes rather than one thousand in each kilobyte of information. A recent attempt to replace the megabyte with a new decimalised unit, the mebibyte, is rubbished by traditionalists who refer to it as a 'maybe byte'.

In the 1980s the American space industry took an equally robust view in the face of a sustained effort by NASA at the start of the international space station programme to convert its suppliers to the metric system. 'When the time came to issue production contracts,' recalled Burt Edelson, a former Associate Administrator of NASA, 'the contractors raised such a hue and cry over the costs and difficulties of conversion that the initiative was dropped. The international partners were unhappy, but their concerns were shunted aside.'

The possible consequences of that decision were not to be appreciated for over a decade, but elsewhere the growing confusion about which units to use produced more immediate results. In 1983 an Air Canada plane in flight from Edmonton to Montreal ran out of fuel and had to glide to an emergency landing in Manitoba after the ground crew had mistakenly measured the fuel payload in pounds instead of kilograms. In Britain there was a tragic outcome involving a newborn baby with heart trouble. The weight of Benjamin Adams was recorded as seven pounds and one ounce, and the doctor treating him was advised to prescribe ten micrograms of the heart drug Digoxin per kilogram of body weight. Translating Imperial weight to metric, the doctor miscalculated by a factor of ten, and ordered a nurse to give Benjamin 320 micrograms of the drug instead of thirty-two. Thirteen hours later the baby died from the overdose. Even though a life was lost, the circumstances – a tired doctor, an urgent case, a snap decision – made it understandable. A subsequent mistake was so meticulously considered, so amazingly expensive, so incredible, it astounded the world.

At precisely 5.01 a.m. on 23 September 1999, NASA's press office issued its first release on the critical phase of its fourth expedition to Mars. The spacecraft, launched over nine months and 416 million miles earlier, was about to enter orbit around the planet, and the release announced that '. . . orbit insertion has begun!' The exclamation mark was a tribute to the intricate engineering and computer work involved, as well as to the approximately $250 million price tag for the probe, including $125 million for the craft and its instruments.

The spacecraft was intended to circle the planet, taking observations of its atmosphere from about sixty-five miles altitude – hence the name Mars Climate Observer. As the engines fired to slow it down, the craft disappeared behind Mars, and ground engineers prepared for its return at 5.26 Pacific Coast Time. It did not reappear on schedule, and the stream of news from the press office grew more sombre in tone as its absence grew longer. Then, late in the day came the first admission that the craft was lost, and an investigation began into the reasons. It took barely ten days to track down the root cause – 'failure to use metric units in the coding of ground software file, "Small Forces"', to quote the deeply embarrassed NASA release.

The clash of two systems had produced a childish mistake. The new team taking over responsibility for the Orbiter at the constructors, Lockheed Martin, was accustomed to dealing in American customary units rather than metric as specified by the Jet Propulsion Laboratory, which was responsible for managing the Mars missions. In writing software for the Orbiter's thruster engine, used to navigate the craft, they specified its force in pounds instead of newtons. (The equation to translate from customary to metric is one pound-force = 0.45359237 kilograms x 9.80665 metres per second squared = 4.448 newtons, approximately.) Thus, each time the thruster fired to stabilise the craft, it was emitting barely one-quarter the force that the instruments on earth recorded. Orbiter was already more than a hundred miles lower in trajectory than it should have been when it went into orbit, and the probability is that it burned up or crashed into the surface of Mars.

Asked for his reaction to the news, Noel Hinners, in charge of flight systems at Lockheed Martin Astronautics, replied with a comment echoed by most of the scientific community: 'The reaction is disbelief. It can't be something that simple that could cause this to happen.'

The fact that it *had* happened provided stark evidence of the state of the United States's measurements. Two hundred years after George Washington pointed out the need for the country to have a single uniform system of weights and measures, it has got two. 'Twenty years ago,' Hinners said in exasperation, 'we went through this whole hassle of, "Should the US go metric?" I wish we had.'

In Britain that point was finally passed on 1 January 2001, when the last relaxation of European Union rules on weights and measures ran out and it became illegal to sell goods by the pound or inch or any other customary measure. The rules allowed market traders to display the old units until 2010, although with their metric equivalent shown more prominently, but after that date uniformity would be everywhere.

If it were a simple question of logic, the United States would also be planning to change. Logically the American customary system is now a museum piece, the last remaining relic of an organic, four-based system that has been preserved across the centuries by a freak of history. It evolved from hands-on, purposeful activity, and every argument that Jefferson put forward about the greater arithmetical simplicity of calculating in tens still holds good. From the moment that the majority of people started to spend most of their lives indoors, and to earn most of their living not from their hands but from mental activity, the advantages of doubling and halving began to dwindle away.

Logic, however, has never played more than a small part in the history of weights and measures. The rest has been about the distribution of power. In its rawest guise, greater accuracy has given empires the power to explore new areas and to exploit them at the expense of the less accurate. But measurement is also about the power of society to allow a just exchange of goods and cash, and at its most fundamental level it has, like language, the power to express a personal value between the individual and the material world. This is what makes the choice faced by the United States so dramatic.

To add to the tension, the European Union has fixed the year 2010 as the deadline after which it will accept only metrically measured goods. Like everything to do with measurement, the EU's decision has a little to do with logic – a single system is more efficient than two – and much to do with control, profits and power. All that has happened in the last two centuries points to an inescapable decision. Eventually the United States government will have to make a choice between economics and its past, between the wishes of business and those of its citizens.

EPILOGUE

The Witness Tree

The California office of the Bureau of Land Management is a white, government-designed box in the outskirts of the city of Sacramento, where Johann Sutter once thought he owned land. It is a peculiarly suitable location for the depository for the survey plats and notes on which all modern real-estate ownership in the state is based. It attracts a wide range of citizens, from suburban homeowners tangling with neighbours over garden boundaries to ranchers pressing for grazing rights, and gold miners establishing claims for a promising outcrop of mountain near Death Valley. If ever there was a place where measurement and ownership come together it is here.

'When you see how easy it is to use the land survey,' declares Lance Bishop, chief of the BLM's geographic services in California, 'you have to admire Thomas Jefferson's foresight in choosing a grid. Every parcel of land has an identity. As an example, I've just bought a five-acre parcel, and I can go back to the original records, and see the shape of the original property, where it was first platted, where the original markers were set, and all subsequent records. It's very clear, there's no ambiguity about what you own.'

It is here too that a preview of the upcoming battle over measurements can be found. At every level except property, surveying has gone metric. The National Geodetic Survey has been so from its origins in 1817, the Geological Survey since 1879, the world's ruling body, La Fédération Internationale Géomètre, since 1865, and now even the Global Positioning

Systems and transits used by every grunt surveyor give measurements in metric units. The figures, however, are translated not just into feet and inches, but chains and links.

'Chain attachment is very strong,' admits Bishop. 'No one knows what it means any more, but no one wants to give it up. We're pushing it – I've tried platting two sections in metric units, but you would not believe the resistance – it is terrific. And you can understand why.'

As early as 1876, the Franklin Institute of Philadelphia listed as a major obstacle to the introduction of the metric system the fact that 'The measurements of every plot of ground in the United States have been made in acres, feet, and inches, and are publicly recorded with the titles to the land according to the record system peculiar to this country.' Other objectors made the same point, that the legal problems of redefining property would be endless. 'Nothing can be more perfect than the land measures of the United States in its square mile sections,' Coleman Sellers, a Cincinnati engineer, declared in 1895, 'all of which would be thrown into endless confusion by the proposed change of the mode of measuring, without the slightest gain to any human being by the operation.'

It need not have been like this. If Jefferson's decimal system had been accepted in one of its forms, the decimal pound and foot would have been at the centre of a system of weights and measures which, backed by American industrial muscle and the universal preference for them as units, might have become the world's favourite system. The historic turn was missed and, left without competition as the only scientifically based, decimalised measurement, the French metre has taken its present place as the simplest, most accurate means of measuring everything between the dimensions of a quark and those of a black hole.

Now the old battle between four and ten has returned, and the arguments from the original eighteenth-century debate have become modern. Yet this time much more is at stake.

Two centuries have so embedded the traditional measures in the land and in the outlook of its inhabitants that they could not be eradicated without the destruction of something irrecoverable. The United States is a democracy built upon a concept of property which owes everything to measurement. If in the long run Joseph Dombey and his copper standards do finally arrive, the values that are lost will have come from the heart of the nation's history.

Six thousand feet up in the Sierras, Ed Patton is in search of history, a corner post hammered into the ground in 1873 by Deputy Surveyor William Minto to mark the point where the south-east corner of Section 36, Township 22 South, Range 36 East, Mount Diablo meridian, met its neighbouring section. Bearded and pony-tailed, Ed is a public land surveyor, employed by the Bureau of Land Management and, as he proudly insists, in direct line of succession from Thomas Hutchins and the men who crossed the Ohio in 1785. He is conducting a re-survey of the area. It is needed because 130 years ago the surveyor running the line north towards Minto's corner belonged to the Benson syndicate, which produced hundreds of false surveys, and the point of intersection marked on his plat owed more to guesswork than bootwork.

This is not commercial land. It is steep mountain-slope made up of dry, stony earth that slides beneath your feet. All that grows up here is sagebrush and stands of pinyon pine whose branches make dark-green clumps against the red-grey rubble. Far down in the valley, however, there is good grazing, and it was the need to delineate where the graziers' property ended that sent Minto and his team across these harsh, magical mountains. The ground is littered with obsidian chips, some shaped into blades and arrowheads by earlier inhabitants who had no need of marks to know that this was their place; but Ed's mind is on his predecessors.

'They ran a straight line through all this, straight forties and eighties [chains], over mountains, canyons, places so steep you need someone to support you while you sight through the transit,' he says admiringly. 'And that survey is the one we all go back to. When you find one of their original corners, it is like a handshake with the past.'

Using Minto's notes and the scars of blazed trees, Ed navigates along a ridge, slithers into a dip, and then, halfway up a rock-covered slope, comes upon a giant, yellow-trunked pinyon reaching out from the past. Carved into the living wood and half-obscured by a ragged rim of bark is a series of runic incisions:

TX͓ IIS RX͓ VI E

SX͓ VI

Ed runs his finger gently around the mark, and translates the runes – Township 22 South, Range 36 East, Section 36. 'That's Minto's blaze,' he says softly. 'When he carved it, this tree would have been no more than twelve inches across. Look at it now. And it'll be here another century.'

This is a witness tree, a record of a human claim to part of nature. In 1942 Robert Frost took the name for the title of a collection of poems about the attachment of people to the land. At the inauguration of President John F. Kennedy in 1961, Frost himself read a poem from the collection, 'The Gift Outright', whose theme revolves around the way that the land moulds those who own it and how the hunger for this particular land made those who felt it American.

> This land was ours before we were the land's.
> She was our land more than a hundred years
> Before we were her people. She was ours
> In Massachusetts, in Virginia,
> But we were England's, still colonials,
> Possessing what we still were unpossessed by,
> Possessed by what we now no more possessed.

Something we were withholding made us weak
Until we found out that it was ourselves
We were withholding from our land of living,
And forthwith found salvation in surrender.
Such as we were we gave ourselves outright
(The deed of gift was many deeds of war)
To the land vaguely realizing westward,
But still unstoried, artless, unenhanced,
Such as she was, such as she would become.

'The Gift Outright', 1942

BIBLIOGRAPHY

Abernethy, Thomas, *Western Lands and the American Revolution* (Boston, 1937)

Adams, John Quincy, *A Report upon weights and measures* (Washington, 1821)

Atran, Scott, *The Cognitive Foundations of Natural History* (Cambridge, 1990)

Aubrey, John, *Brief Lives* (London, 1950)

Beale, John Bordley, *On monies, coins, weights and measures, proposed for the United States of America* (Philadelphia, 1789)

Bedini, Silvio, *Professional Surveyor* magazine 1995–2001

Benese, Richard, *This boke sheweth the maner of measurynge of all maner of lande as woodlande us of lande in the felde and comptynge the true nombre of acres of the same & Newlye invented and compyled Prynted in Southwarke in Saynt Thomas hospytall by me James Nicolson* (London, 1538)

Bureau of Land Management, *Manual of Instructions for the Survey of Public Land* (Washington, 1947)

Burnet, Jacob, *Notes on the Early Settlement of the North-Western Territory* (New York, 1847)

Cajori, Florian, *The Chequered Career of Ferdinand Rudolph Hassler, first Superintendent of the United States Coast Survey. A chapter in the history of science in America* (Boston, 1929)

Cocker, Edward, *Cocker's Decimal Arithmetic* (37th edn, London, 1720)

Conover, Martin, *The General Land Office, its History* (Baltimore, 1923)

Conzen, Michael P. (ed.), *The Making of the American Landscape* (London, 1990)

Digges, Leonard, *A Booke named Tectonicon briefely shewing the exact measuring, and speedie reckoning of all maner of Land, Squares, Timber, Stone, steeples, Pillers, Globes, etc.* (London, 1556)

Dubois, E., *Le Naturaliste Joseph Dombey* (Bourg-en-Bresse, 1934)

Dunaway, Wilma A., 'Speculators and Settler Capitalists' in *Appalachia in the Making: The Mountain South in the Nineteenth Century* (North Carolina, 1995)

Durham, Philip (ed.), *The Frontier in American Literature* (New York, 1969)

Elkins, Stanley and MacKitrick, Eric, *The Age of Federalism* (Oxford, 1995)

Elton, G.R., *England under the Tudors* (London, 1974)

Ferguson, Kitty, *Measuring the Universe* (New York, 1999)

Fitzherbert, John, *The Art of Husbandry* (London, 1523)

Frängsmyr, Tore, Heilbron, J.L. and Rider, Robin, etc., *The Quantifying Spirit of the Eighteenth Century* (Berkeley, 1990)

Gibson, Robert, *A Treatise of Practical Surveying: with alterations and amendments adapted to the use of American surveyors* (7th edn, Philadelphia, 1796)

Gunter, Edmund, *Canon triangularum* (London, 1620)

Gunter, Edmund, *The First Book of the Cross-Staffe* (London, 1620)

Gunter, Edmund, *New Projection of the Sphere* (London, 1623)

Gunter, Edmund, *Description and Use of the Sector, the Crosse-staffe and other Instruments* (London, 1624)

Hamy, E., *Joseph Dombey* (Paris, 1905)

Hart, John Fraser, *The Land that Feeds Us* (New York, 1991)

Hassler, Ferdinand Rudolph, *Report on Weights and Measures: Comparison of Weights and Measures of length and capacity* (report to the Senate of the United States by the Treasury Department, 1832)

Hassler, Ferdinand Rudolph, *Coast survey of the United States. [An answer to certain statements on that subject made in Congress]* (Philadelphia, 1842)

Hicks, Frederick C. (ed.), *Thomas Hutchins: A Topographical*

Description of Virginia, Pennsylvania, Maryland and North Carolina (Cleveland, 1778)

Hurt, R. Douglas, *American Farms: Exploring their History* (Krieger Publishing, 1996)

Jackson, John Brinckerhoff, *American Space* (New York, 1972)

Jacobs, Harvey M. (ed.), *Who Owns America? Social Conflict over Property Rights* (Wisconsin, 1998)

Jefferson, Thomas (Boyd, ed.), *The Papers of Thomas Jefferson* (Princeton, 1950–92)

Johnson, Hildegard Binder, 'The United States Land Survey as a Principle of Order', in *Pattern and Process: Research in Historical Geography* (Howard University Press, 1975)

Johnson, Hildegard Binder, *Order Upon the Land: The US Rectangular Survey and the Upper Mississippi Country* (London, 1976)

Johnson, Hildegard Binder, 'The Orderly Landscape' (lecture, Minneapolis, 1977)

Keay, John, *The Great Arc* (London, 2000)

Klein, Arthur, *The World of Measurements* (London, 1975)

Koch, Adrienne, *Jefferson and Madison* (University Press of America, 1986)

Kula, Witold (trans. R. Szreter), *Measures and Men* (Princeton, 1986)

Langewiesche, Wolfgang, in *Harper's Magazine*, October 1950, pp.176–98

Leybourn, William, *The Compleat Surveyor* (London, 1653)

Livermore, Shaw, *Early American Land Companies and their Influence: A Corporate Development* (Commonwealth Fund, 1939)

Love, John, *Geodaesia or the art of surveying and measuring of land made easie* (London, 1688)

Marryat, Thomas, *A Diary in America* (New York, 1962)

Meinig, D.W., *The Shaping of America* (New Haven, 1986, 1993, 1998)

Melish, John, *Travels in the United States 1806, 1807 & 1809* (Philadelphia, 1813)

Merrell, James, *Into the American Woods: Negotiations on the Pennsylvania Frontier* (London, 1999)

Nash, *Wilderness and the American Mind* (New Haven, 1967)

National Institute of Standards and Technology, *Toward a Metric America* (Washington, 1971)

Newman, James, *The World of Mathematics* (New York, 1956)

Pasley, Jeffrey, 'Private Access and Public Power', in *The House and the Senate in the 1790s: Petitioning, Lobbying, and Institutional Development* (Ohio University Press, 2002)

Pattison, W.D., *The Beginning of the American Rectangular Land Survey System* (Ohio Historical Society, 1970)

Pattison, W.D., *Origins of the American Rectangular Land Survey System 1784–1800* (University of Chicago Department of Geography Research Paper No.50, 1957)

Price, Edward T., *Dividing the Land: Early American Beginnings of our Private Property Mosaic* (University of Chicago, 1995)

Professional Surveyor magazine 1995–2001

Putnam, Rufus (ed. Rowena Buell), *Memoirs of Rufus Putnam* (New York, 1904)

Pynchon, Thomas, *Mason & Dixon* (New York, 1997)

Raban, Jonathan, *Bad Land: An American Romance* (London, 1996)

Reps, John, *The Making of Urban America: A History of City Planning in the United States* (Princeton University Press, 1965)

Richeson, A.W., *English Land Measuring* (Cambridge, Mass., 1966)

Sakolski, A.M., *The Great American Land Bubble: The Amazing Story of Land-Grabbing, Speculations, and Booms from Colonial Days to the Present Time* (New York, 1932)

Salar, J. Riquelme, 'El doctor Dombey' in *Proceedings of the International Congress of the History of Medicine* (1959), pp.160–2

Schama, Simon, *Landscape and Memory* (London, 1995)

Selby, John, *The Conquest of the American West* (London, 1975)

Smalley, E.V., 'Isolation of Life on Prairie Farms', in *Atlantic Monthly*, 1893

Smith, Thomas H., *The Mapping of Ohio* (Kent State University Press, 1977)

Steele, A.R., *Flowers for the King: The Expedition of Ruiz and Pavon and the Flora of Peru* (Durham, North Carolina, 1964)

Stilgoe, John, *The Common Landscape of America 1580–1845* (New Haven, 1982)

Stilgoe, John, *Borderland: Origins of the American Suburb 1820–90* (New Haven, 1988)

Taylor, E.G.R., *Mathematical Practitioners of the Tudor and Stuart Period* (London, 1954)

Thompson, F.M.L., *Chartered Surveyors: The Growth of a Profession* (London, 1968)

Trollope, Frances, *Domestic Manners of the Americans* (London, 1969)

Turner, F.J., *The Frontier in American History* (London, 1996)

United States Congress, Journals of the American Congress 1774–88 (Washington D.C., 1904)

Wallace, Anthony F.C., *Jefferson and the Indians: The Tragic Fate of the First Americans* (Harvard, 1999)

White, C. Albert, *A History of the Rectangular Survey System* (Government Printing Office, Washington, 1982)

Williams, Penry, *Life in Tudor England* (London, 1964)

Wilson, Henry, *Geodesia catenea or surveying by the chain only* (London, 1786)

Winstanley, Gerrard, *A DECLARATION FROM THE Poor oppressed People OF ENGLAND, DIRECTED To all that call themselves, or are called Lords of Manors, through this NATION; That have begun to cut, or that through fear and covetousness, do intend to cut down the Woods and Trees that grow upon the Commons and Waste Land* (London, 1649)

Winstanley, Gerrard, *A LETTER TO The Lord Fairfax, AND His Councell of War, WITH Divers Questions to the Lawyers, and Ministers: Proving it an undeniable Equity, That the common People ought to dig, plow, plant and dwell upon the Commons, without hiring them, or paying Rent to any* (London, 1649)

Wise, M. Norton, *The Values of Precision* (Princeton, 1995)

Worsop, Edward, *A discoverie of sundrie errours and faults daily*

committed by landemeters ignorant of arithrmeticke and geometrie
(London, 1582)

Xenophon (trans. Gentian Nervet), *Oeconomicon* (London, 1534)

Zengerle, Jason, 'Waits and Measures', in *Mother Jones*, January/February 1999

Zupko, Ronald Edward, *Revolution in Measurement: Western Euopean Weights and Measures Since the Age of Science* (Philadelphia, 1990)

INDEX

Biot, Jean-Baptiste: *Essai sur l'histoire générale des sciences* 138
Bird, John: Troy pound 217; yard 128, 217; zenith sector 28
Birkbeck, Morris 188
Bishop, Lance 287, 288
Bismarck, Prince Otto von 265, 276
Bloch, Marc 4
Blodget, Lorin: *Climatology* 247, 249
Blount, William 167, 170
Blue Ridge mountains 25, 27, 39, 47, 49
Board of Trade, London 47
Boone, Daniel 38, 40, 45
Borda, Jean-Charles 101, 129–30, 137–8, 140, 142–3, 266, 271
Bordley, John Beale 105; *On monies, coins, weights . . .* 118
Boston, Massachusetts 23, 42, 48; brewers 65
Boston Gazette 120
Bouchon, Monsieur (Louisiana Surveyor-General) 213
Boulton, Matthew 127
bowshots 17
Boyd, Julian 118
Bradford, William, Governor of Plymouth colony 32
brewers' measurements 63
Bridgewater, Earl of 10
Briggs, Isaac 171–2, 173
Britain: and America 23, 24, 33, 34–5, 37–8, 41, 42–4, 114, 155, 206, 226; and American War of Independence 47–8, 57; Civil War 37; Crown lands 13–14; currency 20–1; dissolution of the monasteries 16–17; enclosures 3, 16–17, 23, 37; instrument-makers 206; mapping of 3, 4–5, 13, 128, 130–1, 204–5; wars with France 139, 148–9; weights and measures 20, 21–3, 110, 124–8, 131–2, 217, 221, 275–8, 281, 282, 283, 285; *see also* Scotland
British Army: Board of Ordnance 131
British Association for the Advancement of Science 267, 268
Brittany, France 91
Brown, Captain Nathaniel Williams 147–8, 150, 151–2
Brown, Uria 166, 168
buffalo 254, 257
Bureau of Land Management 255, 287, 289
Burlington Railroad Company 199, 200
Burnham, Daniel 200–1
Burr, Jeheu 30
Burroughs, John 254
Burt, William A. 223–5, 229–31, 245
bushels 17, 18, 21, 113, 125, 127; France 261; New York 214; Pennsylvania 104; Winchester 125, 221
Byrd, William, II 49, 179; *The History of the Dividing Line* 25–6, 35, 51

Cabot, Sebastian 4
California 225, 238, 239, 240, 242–4, 256; border 243, 256; gold rush 191, 201; under Spanish rule 33–4
Calverts of Maryland 28
Camden, Charles Pratt, 1st Earl 45
Canada 34, 228–9, 273, 277; establishment of frontier 208; metrication 281–2, 283; *see also* Montreal; Quebec
Canadian Pacific Railroad 250
Carnegie, Andrew 260

Henry III, of England 20
Henry VII, of England 21
Henry VIII, of England 4, 6, 23
Henry, Patrick 45, 164, 169–70
hides 5
Hill, James J. 248–9
Hinners, Noel 285
hogsheads 64–5, 127
Holme, Thomas: Philadelphia 197
Homestead Act (1862) 247, 248
homesteaders 250–4, 255–8
Hopewell mounds 84, 85
Hopkinson, Francis 69
horse-belly view 17–18
Horton, W.F.: *Landbuyer's, Settler's and Explorer's Guide* 184
houpées 17
Howell, David 57
Hume, David 43, 115
Huntington, General 59
Huntsville Republican 185–6
Hutchins, Thomas xviii–xix, 71–9; map 39–40

Idaho–Montana border 246
Illinois 45, 67, 156, 162, 178, 179, 188, 201, 212, 250, 281; Sangamon County 180, 236
Illinois and Wabash Company 45
Imperial units 220–1
Imperial Weights and Measures Act (1824) 226
inch 280–1, 277
India 226; measurements 19; metrication 273–4
Indian Removal Act (1830) 234–5
Indiana 45, 71, 156, 162, 179, 212; border 177
Indianapolis, Indiana 196, 199
Indians 27, 44, 61, 141, 231–7; Apache 241; Cherokee 40, 43–4, 170, 172, 185, 233, 235; Chickasaw 170, 185, 233, 235; Chippewa 229; Choctaw 170,

185, 233, 235; Comanche 241; Creek 233, 235; Delaware xviii, 76, 84, 119, 155, 232, 234; Duwanish 235; Iroquois xix, 34, 76; Kaskaskia 120, 232; Kickapoo 236, 248; Miami xix, 76, 119, 120, 155, 156; Nez Percé 236; Ojibwe 229, 231–2; Ottawa 76; Plains 17–18; Powhatan 33, 34; Seminole 233, 235; Seneca 234; Shawnee 40, 76, 234; Sioux 257; Six Nations 34, 40, 119; Tuscarawas 44; Wampanog 34; Wyandot 119, 155, 156, 234
Industrial Revolution 127, 260
International Geodetic Association 269
International Office of Weights and Measures 269, 270, 271
International System of Weights and Measures 271
Iowa 177, 184, 245, 250, 281
Irish settlers 35, 103, 114, 179
Iroquois Indians xix, 34, 76
Italy 88, 266
Ives, William 225, 235
Izard, Ralph 123, 140

Jackson, President Andrew 190, 214, 221–2, 233
Jackson County, Missouri 197–8
Jacobins 134, 139, 146, 260
James I, of England: 1609 charter 24
Jamestown colony, Virginia 32–3
Japan: metrication 274
Jay Treaty 155
Jefferson, Martha Wayles Skelton 49, 56–7
Jefferson, Peter 27, 38, 45, 49, 54
Jefferson, Thomas: character 48, 51–2, 62; and currency 62,